W9-CXL-762

The
Princeton
Review

Cracking the AP Biology Exam

By Kim Magloire

2004–2005 Edition

Random House, Inc.
New York

www.PrincetonReview.com

The Independent Education Consultants Association recognizes The Princeton Review as a valuable resource for high school and college students applying to college and graduate school.

Princeton Review Publishing, L.L.C.
2315 Broadway
New York, NY 10024
E-mail: booksupport@review.com

ISBN: 0-375-76393-7
ISSN: 1092-0080

*AP and Advanced Placement are registered trademarks of the College Entrance Examination Board, which was not involved in the production of, and does not endorse, this book.

Editor: Allegra Viner
Production Editor: Jodie Gaudet
Production Coordinator: Ryan Tozzi

Manufactured in the United States of America.

9 8 7 6 5 4 3 2

2004–2005 Edition

ACKNOWLEDGMENTS

Special thanks go to Sean Barry, Allegra Viner, Judene Wright, Patricia Dublin, Ryan Tozzi, John Bergdahl, Maria Dente, Jose Freire, Effie Hadjiioannou, Jason Kantor, Stephanie Martin, Len Small, and Amy Zavatto. Finally, I would like to dedicate this book to my parents who instilled in me a love for science.

BRIEF CONTENTS

CONTENTS

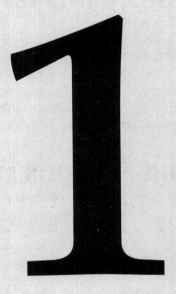

1

Introduction

WHAT IS THE AP BIOLOGY EXAM?

So you've just spent the better part of a year in an "advanced placement" biology course. And what have you learned? Biology, sure. But what kind of biology?

In theory, you're midway through the equivalent of a college-level biology course. However, high school courses vary, to say the least. Sometimes you get a great teacher, sometimes . . . you don't. The Advanced Placement Biology Exam is a way to determine if the course you've taken is up to par: Have you really learned a year's worth of college-level biology?

That's what the AP test is intended to measure. You take it so that colleges can determine if you've mastered the material that the average college freshman learns in his or her introductory biology course. If so, you'll be eligible for college credit, advanced standing, or both, depending on the college. Different colleges have different policies, so make sure you find out from the colleges you intend to apply to exactly what their policies are.

What if you're not enrolled in an AP course? Provided you've seen the same topics as those who are enrolled in such a course, there's no reason why you shouldn't consider taking the AP Biology Exam. However, before you sign up, check with your biology teacher. He or she is in a much better position to determine if you've actually done college-level science this past year.

WHO WRITES THE AP BIOLOGY EXAM?

The AP Biology Exam is written by the College Board, which is a division of ETS, the **Educational Testing Service**. These are the same folks who bring you all your standardized tests, from the PSAT and the SAT to the rest of your SAT II subject tests. How do they go about it?

The College Board puts together a committee of university and high school teachers. This committee determines the content and format of the AP exam. Why should you care about ETS and the College Board? Well, you shouldn't. But you should care about how they write the test, which is extremely important when it comes to our approach.

THE PRINCETON REVIEW APPROACH

There are basically two ways to prepare for the AP Biology Exam:

- *Know absolutely everything about everything.* This is ETS's way. Bad idea.

- *Review only what you need to know, and tackle the test strategically.* This is The Princeton Review's way—and the best way to improve your score.

Rather than trying to teach you everything there is to know about biology, we at The Princeton Review focus on test-taking strategies. Naturally, we'll review some hard science as well. But rather than cluttering your brain with dull lectures and obscure terms, we'll look at only the biology you need to know for the test, explaining and highlighting key concepts along the way. But who are we and how do we know so much about what's important for the AP Biology Exam?

The Princeton Review is the nation's fastest growing test preparation company. We've been at it for more than 25 years, preparing students for standardized tests by showing them how to beat ETS at their own game. Our insight into the AP Biology Exam is the fruit of intensive analysis of heaps of AP exams. For you, this translates into a relatively painless, sure-fire approach to beefing up your AP score.

In this book, we'll show you how best to take the AP Biology Exam because we know exactly how it's put together. By understanding how the test is written, we'll be able to help you outfox the test writers in two ways:

- By reviewing only the biology you need to know for the test

- By giving you simple, straightforward strategies for answering multiple-choice questions and for writing essays

By the time you finish this book, you'll have both the science and the strategies you'll need to beat the AP Biology Exam.

HOW MUCH BIOLOGY DO YOU NEED TO KNOW?

Fortunately, we've already done the groundwork for you. We know exactly what ETS likes to test and how they test it. The College Board has put together a list of the topics covered on the AP Biology Exam, as well as a breakdown of the frequency with which they appear on the test.

The AP Biology Exam covers three major areas:

- Molecules and Cells

- Heredity and Evolution

- Organisms and Populations

These three areas are further subdivided into major topics. By the way, the percentages given below will give you a rough idea of the percentage of questions from each category that will appear on the test. For instance, since 10 percent of the test concerns cells (see below), you can expect that about 10 percent of the multiple-choice questions—10 questions altogether—will deal with cells.

Here, then, is the breakdown:

1. Molecules and Cells (25%)
    A. Chemistry of Life (7%)
        • Organic molecules in organisms
        • Water
        • Free-energy changes
        • Enzymes
    B. Cells (10%)
        • Prokaryotic and eukaryotic cells
        • Membranes
        • Subcellular organization
        • Cell cycle and its regulation
    C. Cellular Energetics (8%)
        • Coupled reactions
        • Fermentation and cellular respiration
        • Photosynthesis
2. Heredity and Evolution (25%)
    A. Heredity (8–9%)
        • Meiosis and gametogenesis
        • Eukaryotic chromosomes
        • Inheritance patterns
    B. Molecular Genetics (8%)
        • RNA and DNA structure and function
        • Gene regulation
        • Mutation
        • Viral structure and replication
        • Nucleic acid technology and applications
    C. Evolutionary Biology (8%)
        • Early evolution of life
        • Evidence for evolution
        • Mechanism of evolution

3. Organisms and Populations (50%)

   A. Diversity of Organisms (8%)
      - Evolutionary patterns
      - Survey of the diversity of life
      - Phylogenetic classification
      - Evolutionary relationships

   B. Structure and Function of Plants and Animals (32%)
      - Reproduction, growth, and development
      - Structural, physiological, and behavioral adaptation
      - Response to the environment

   C. Ecology (10%)
      - Population dynamics
      - Communities and ecosystems
      - Global issues

In addition to the outline above, we have recently seen a few minor content-related adjustments to the examination. Although we have already included the following content in our comprehensive subject review, it has not previously been seen in questions on the AP Biology Exam. You may now see questions requiring familiarity with representative organisms from the three domains: Bacteria, Archaea, and Eukarya. In addition, more emphasis may be placed on identifying the distinguishing characteristics of each group of phylogenetic classification (domains, kingdoms, and the major phyla and divisions of animals and plants). Other than these minor updates to material tested, there are no major changes in the overall concepts. Furthermore, the distribution of questions relating to each of the topics taught in the course also remains unaffected.

This might seem like an awful lot of information. But for each topic, there are only a few key facts you'll need to know. Your biology textbooks go into far greater detail about each of these topics than we do. That's because they're trying to teach you "correct science," whereas we're aiming to improve your scores. Our science is perfectly sound, it's just cut down to size to save you some brain room. We've done away with the details and given you only what's important. Moreover, as you'll soon see, our treatment of these topics is far easier to handle.

THE STRUCTURE OF THE TEST

The AP Biology Exam is three hours long and is divided into two sections: Section I (multiple-choice questions) and Section II (essay questions).

Section I consists of 100 multiple-choice questions. These are further broken down into three parts: (1) regular multiple-choice questions, (2) matching questions, and (3) questions dealing with experiments or data. ETS gives you 80 minutes to answer all 100 of these questions.

Section II involves essays. You'll be presented with four essay questions touching upon key issues in biology. You're given a 10-minute reading period followed by 90 minutes to answer all four essays.

If you're thinking that this sounds like a heap of work to try to finish in three hours, you're absolutely right. Here's how it breaks down: You have roughly 45 seconds per multiple-choice question and 22 minutes per essay. Gulp. How can you possibly tackle so much science in so little time?

Fortunately, there's absolutely no need to. As you'll soon see, we're going to ask you to leave a small chunk of the test blank. Which part? The parts you don't like. This selective approach to the test, which we call "pacing," is probably the most important part of our overall strategy. But before we talk strategy, let's look at the different types of questions in the first section of the test.

Section 1

There are basically three parts to the first section:

- Part 1—contains regular multiple-choice questions
- Part 2—is made up of matching questions
- Part 3—consists of multiple-choice questions dealing with an experiment or a set of data

Part 1

Part 1 of the AP Biology Exam consists of about 58 run-of-the-mill multiple-choice questions. These questions test your grasp of the fundamentals of biology (or so ETS likes to think). Here's an example:

22. If a segment of DNA reads 5´-ATG-CCA-GCT-3´, the mRNA strand that results from the transcription of this segment will be

(A) 3´-TAC-GGT-CGA-5´
(B) 3´-UAC-AGT-CAA-5´
(C) 3´-TAA-GGU-CGA-5´
(D) 3´-TAC-GGT-CTA-5´
(E) 3´-UAC-GGU-CGA-5´

Don't worry about the answer to this question for now. By the end of this book, it will be a piece of cake. The majority of the questions in part 1 are presented in this format. A few questions may include a figure, a diagram, or a chart.

Part 2

The second part is slightly different. In part 2, you are asked to match lettered portions of a diagram or a list to numbered statements. This "matching" exercise usually tests your knowledge of different structures. Let's look at an example:

Questions 63–66

(A) Thyroid
(B) Parathyroid
(C) Hypothalamus
(D) Pancreas
(E) Adrenal cortex

63. Secretes hormones that regulate plasma glucose levels

64. Secretes aldosterone

65. Regulates the basal metabolic rate in the body

66. Secretes gonadotrophic releasing and inhibiting factors

Again, don't worry if you don't know the answers yet. We just want to show you what the questions look like. You're asked to match the numbered questions to the lettered organs above. Sometimes you'll be given a group of questions, all of which refer back to a diagram or an illustration. You'll see six to seven of these groupings on the test.

Part 3

The last part of the first section also consists of multiple-choice questions, yet here you're asked to think logically about different biological experiments or data. Here's a typical example:

Questions 99 and 100 refer to the following graph and information.

To understand the workings of neurons, an experiment was conducted to study the neural pathway of a reflex arc in frogs. A diagram of a reflex arc is given below.

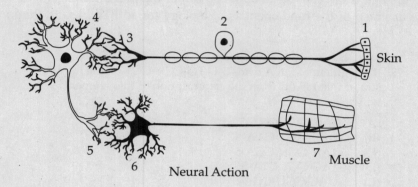

Neural Action

99. Which of the following represents the correct pathway taken by a nerve impulse as it travels from the spinal cord to effector cells?

(A) 1-2-3-4
(B) 6-5-4-3
(C) 2-3-4-5
(D) 4-5-6-7
(E) 7-6-5-4

100. The brain of the frog is destroyed. A piece of acid-soaked paper is applied to the frog's skin. Every time the piece of paper is placed on its skin, one leg moves upward. Which of the following conclusions is best supported by the experiment?

(A) Reflex actions are not automatic.
(B) Some reflex actions can be inhibited or facilitated.
(C) All behaviors in frogs are primarily reflex responses.
(D) This reflex action bypasses the central nervous system.
(E) Reflex responses account for a large part of the total behavior in frogs.

You'll notice that these particular questions refer to an experiment. In part 3, the experiments, like this one, usually require a little more than just the basics of biology. ETS claims that the questions in part 3 test your ability to integrate information, interpret data, and draw conclusions from the results. More often than not, common sense and logical deduction are a lot more effective here than is "strict science."

SECTION II

This section of the test consists of four free-response questions—the essays. The questions are usually divided into parts and vary in difficulty. In this section, the test writers are testing your grasp of the major concepts and themes in biology rather than your ability to memorize facts.

Remember the three major areas we said ETS likes to test? Well, one question will come from area 1, Molecules and Cells, another question will come from area 2, Heredity and Evolution, and two questions will come from area 3, Organisms and Populations. The trend on the AP test is to focus on themes that crop up time and again in a typical AP biology course.

Take a look at the themes put together by the AP Biology Development Committee:

- Science as a Process

- Evolution

- Energy Transfer

- Continuity and Change

- Relationship of Structure to Function

- Regulation

- Interdependence in Nature

- Science, Technology, and Society

Now let's look at a sample free-response or essay question:

1. Enzymes are biological catalysts.

   a. Relate the chemical structure of an enzyme to its catalytic activity and specificity.
   b. Design an experiment that investigates the influence of temperature, substrate concentration, or pH on the activity of an enzyme.
   c. Describe what information concerning enzyme structure could be inferred from the experiment you have designed.

Notice that this question has three parts. Part *a* tests one of the themes from the list: the relationship between structure and function. Part *b* asks you to design an experiment based on your knowledge of enzymes. Part *c* refers back to your experimental design and asks you to interpret your results and draw inferences from it. This section tests not only your knowledge of biology but also your organizational and writing skills.

Now that we've had a peek at the science you'll need to know and the structure of the test, let's look at the strategies that will help you ace the AP Biology Exam.

USING THE PRINCETON REVIEW APPROACH TO CRACK THE SYSTEM

We mentioned earlier that our approach is strategy-based. As you're about to see, many of these strategies are based on common sense—for example, using mnemonics like "ROY G. BIV." (Remember that one? It's the mnemonic for red, orange, yellow, green, blue, indigo, violet—the colors of the spectrum.) Others are not so common-sensical. In fact, we're going to ask you to throw out much of what you've been taught when it comes to taking standardized tests.

There are eight strategies that we'll ask you to apply come test time:

- Strategy 1: Pace Yourself
- Strategy 2: The Three-Pass System
- Strategy 3: Process of Elimination
- Strategy 4: Aggressive Guessing
- Strategy 5: Word Associations
- Strategy 6: Mnemonics
- Strategy 7: Identify Question Types
- Strategy 8: The Art of the ETS Essay

Let's take a look at the Princeton Review approach.

STRATEGY 1: PACE YOURSELF

When you take a test in school, how many questions do you answer? Naturally, you try to answer all of them. You do this for two reasons: (1) Your teacher told you to, and (2) if you left a question blank, your teacher would mark it wrong. However, that's not the case when it comes to the AP Biology Exam. In fact, finishing the test is the worst thing you can do. Before we explain why, let's talk about timing.

One of the main reasons that taking the AP Biology Exam is so stressful is the time constraint we discussed above—45 seconds per multiple-choice question and 22 minutes per essay. If you had all day, you would probably do much better. We can't give you all day, but we can do the next best thing: We can give you more time for each question. How? By having you slow down and answer fewer questions.

Slowing down, and doing well on the questions you do answer, is the best way to improve your score on the AP Biology Exam. Rushing through questions in order to finish, on the other hand, will always hurt your score. When you rush, you're far more likely to make careless errors, misread, and fall into traps. Keep in mind that for every wrong answer choice you pick in Section I, you lose one-quarter of a point. Blank answers, on the other hand, are not counted against you.

By now you're asking yourself, "How do they know this works?" Don't take our word for it. We'll walk you through an example to prove our point. But before we do so, let's take a look at how the AP Biology Exam is scored.

The AP translation game

The maximum number of points you can earn on the AP Biology Exam is 100 points for the multiple-choice questions in Section I and 40 points for the four essay questions in Section II. These "raw scores" are translated to "composite scores." ETS has set up the test so that Section I, with its 100 questions, counts for 60 percent of your overall grade, while Section II, with its four essays, counts for only 40 percent of your grade. These composite scores are then further translated to numbered grades ranging from 1 to 5. Here's how it's done.

For Section I, ETS takes the number of questions you answered incorrectly, divides it by four, and subtracts the result from the number of questions you answered correctly:

(Raw score for Section I) = (Number answered correctly) – (Number wrong ÷ 4)

Next, ETS takes that raw score and converts it to a composite score by multiplying it by 0.75:

(Composite score for Section I) = 0.75 × (Raw score for Section I)

If you got every question right on this portion of the test, you would have a raw score of 100. The highest composite score, therefore, is 75 (i.e., 100 × 0.75 = 75).

For Section II, you can earn up to 10 points for each essay question for a total of 40 points:

(Raw score for Section II) = (Points for Essay 1) + (Points for Essay 2) + (Points for Essay 3) + (Points for Essay 4)

Your raw score is then multiplied by 1.5 to yield a composite score:

(Composite score for Section II) = 1.5 × (Raw score for Section II)

If you wrote perfect essays, you would get the perfect raw score of 40. The highest number of points you can get, therefore, would be 60 (i.e., 40 × 1.5 = 60).

Remember that on Section I, our maximum was 75 points. Combined with the maximum total of 60 for Section II, we get a combined maximum of 135 points. ETS adds up the total composite scores for both sections and converts their sum into a simple, single-digit grade:

THE ETS GRADING SYSTEM		
Composite Score	**AP Grade**	**Comment**
82–135	5	Extremely well qualified
63–81	4	Well qualified
47–62	3	Qualified
29–46	2	Possibly qualified
0–28	1	No recommendation

Seems terribly complicated, doesn't it? Fortunately, you don't need to memorize how ETS computes your score. You only need to know how to apply these conversions when determining your score on the practice test.

What's a decent score? Naturally, you'd like to get a 3 or better. Most schools will accept a score of 3 or above as equivalent to a year of college biology. According to past exams, roughly two-thirds of the students who take the AP Biology Exam receive a grade of 3 or above. This means that in order to "pass" this test, you need to be somewhere in that top two-thirds. What do you need to do in order to get there?

Ms. C. Darwin takes an AP test

Suppose that a random student—we'll call her C. Darwin—took the AP Biology Exam and left half of Section I blank. That means she did 50 questions and left out 50 questions. Let's say that out of the 50 she did, she missed 20. Keep in mind that you lose one-quarter point for each wrong answer. That gives Ms. Darwin:

$$37 \text{ right} - (20 \text{ wrong} \div 4) = 32 \text{ raw points}$$

This may not sound like a great performance. On a regular test, it would be an all-out flop, wouldn't it? A score of 40 points out of 120 is not the kind of grade Ms. Darwin is likely to stick on the refrigerator door. After all, it's only 33 percent. Let's see what happens here.

Her raw score for Section I would have been a 40. We multiply this raw score by 0.75 to get a composite score:

$$0.75 \times 32 = 24$$

For Section II, let's assume she answered all four essays and got 5 points for each question—once again, that's only half the total potential! Altogether, that gives us a total of 20 points for Section II.

To obtain the composite score for Section II, we multiply the raw score by 1.5:

$$1.5 \times 20 = 30$$

Taking the two composite scores, we can figure out Ms. Darwin's total composite score:

Section I: 24 points

Section II: 30 points

Total: 54 points

Using the table above, you can see that a total composite score of 54 translates to a grade of 3 on the AP Biology Exam. Take a look back. Ms. Darwin got a 3, a passing grade, even though she blew off half of Section I and received only half credit on the essays! Not bad!

What all this means for you

The bottom line is this: There's absolutely no reason to finish the test, even if you're shooting for a score higher than a 3. This is particularly true of Section I, though we'll soon see how the same thinking applies to Section II. It may be surprising to think of test taking in this way, but it really works. By the way, the conversions we've provided are based on ETS's own calculations, the same ones they provide in their materials. Why don't they let you know that you can skip half the test and still get a decent grade? Hmmm....

But what if you're shooting for something higher than a 3? To reach that goal, you may have to answer a few more questions—but not many more. Even if you're aiming for a 4 or a 5, there's absolutely no reason to finish the test. Simply slow down so that you can do better on the questions you're comfortable with. Which ones are those? Obviously, the ones you know!

Let's take a look at precisely how many questions you need to get the score you desire:

THE AP BIOLOGY PACING CHART		
To Get This Score:	Do This Many Questions:	
	Section I	Section II
1	25	2 to 3 essays
2	40	3 essays
3	60	3 to 4 essays
4	85	4 essays
5	100	4 essays

But what if you don't nail every question in your range? Don't sweat it. We've already figured your mistakes into the pacing chart. We've assumed that, being human, you'll miss a few questions on Section I and probably miss some of the essay points on Section II. Take a look back at our example above.

Our imaginary student, Ms. Darwin, aiming for a 3, left half of Section I blank, and still missed about two-fifths of the questions she did. Nevertheless, with a decent performance on the essays (half credit), she still landed comfortably in the 3 range. This boils down to a very simple point, one that we like to think of as the golden rule of pacing:

Do fewer questions and your score will immediately improve.

Even if you are aiming for a 5, there's no reason to answer all the questions. Remember that the 5 range, the highest score on the exam, starts at a composite of about 82. This is only 60 percent of the total possible score of 135. Hardly what you'd think of as a "perfect" grade.

For us, this means that there's no reason to get bogged down with extremely difficult problems. If you come across a question that completely stumps you, skip it! You don't need it anyway, not even to get a perfect score. If, however, you spend lots of time on very tough problems, you're less likely to have time for the problems you do know and consequently less likely to get the score you desire.

Taking an actual test

If you have not already taken a practice test, take one now. It will give you a much clearer idea of where you are in terms of pacing. It's a good idea to take your first practice test the "regular" way, that is, without the pacing chart. This will give you an idea of what it's like to try to do 100 multiple-choice questions and four essays in only three hours. After you've completed our book, go back and take a second practice test, applying the new techniques and biology review you've gained from this book.

We've provided you with two practice tests in the back of the book. We also recommend that you purchase an AP Biology Exam from ETS. You can do these tests in any order you like. If you do not have an ETS test, you might want to start with ours and do the ETS test once you've acquired it.

After you've scored your first test using the guidelines spelled out above, you'll know how many questions you'll need to answer to reach your goal. Pace yourself wisely and you're already on the path to higher scores.

STRATEGY 2: THE THREE-PASS SYSTEM

According to the pacing chart, even those who want a perfect score do not have to answer all the questions on the test. The rest of us have even more leeway: We can leave up to half the test blank and get a 3, as we saw from Ms. Darwin's test. But which questions should we skip? The answer is pretty simple:

Skip the questions you don't like.

The AP Biology Exam covers a broad range of topics. There's no way, even with our extensive review, that you will know everything about every topic in biology. What then should you do?

Do the easiest questions first

The best way to rack up points is to focus on the easiest questions first. Many of the questions asked on the test will be straightforward and require little effort. If you know the answer, nail it and move on. Others, however, will not be presented in such a clear, simple way. As you read each question, decide if it's easy, medium, or hard. During a first pass, do all the easy questions. If you come across a problem that seems time-consuming or completely incomprehensible, skip it. Remember:

Easier questions count just as much as harder ones, so your time is better spent on shorter, easier questions.

Save the medium questions for the second pass. These questions are either time-consuming or require that you analyze all the answer choices (i.e., the correct answer doesn't pop off the page). If you come across a question that makes no sense from the outset, save it for the last pass. You're far more likely to fall into a trap or settle on a silly answer.

Watch out for those bubbles!

Since you're skipping problems, you need to keep careful track of the bubbles on your answer sheet. One way to accomplish this is by answering all the questions on a page and then transferring your choices to the answer sheet. If you prefer to enter them one by one, make sure you double-check the number beside the ovals before filling them in. We'd hate to see you lose points because you forgot to skip a bubble!

So then, what about the questions we don't skip?

STRATEGY 3: PROCESS OF ELIMINATION (POE)

On most tests, you need to know your material backward and forward to get the right answer. In other words, if you don't know the answer beforehand, you probably won't answer the question correctly. This is particularly true of fill-in-the-blank and essay questions. We're taught to think that the only way to get a question right is by knowing the answer. However, that's not the case on Section I of the AP Biology Exam. You can get a perfect score on this portion of the test without knowing a single right answer…provided you know all the wrong answers!

What are we talking about? This is perhaps the single most important technique in terms of the Multiple-Choice section of the exam. Let's take a look at an example on the next page:

1. The structures that act as the sites of gas exchange in a woody stem are the

   (A) lungs
   (B) gills
   (C) lenticels
   (D) ganglia
   (E) lentil beans

Now if this were a fill-in-the-blank-style question, you might be in a heap of trouble. But let's take a look at what we've got. You see "woody stem" in the question, which leads you to conclude that we're talking about plants. Right away, you know the answer is not (A), (B), or (D) because plants don't have lungs, gills, or ganglia. Now we've got it down to (C) and (E). Notice that (C) and (E) are very similar. Obviously, one of them is a trap. At this point, if you don't know what "lentil beans" are, you have to guess. However, even if we don't know precisely what they are, it's safe to say that most of us know that lentil beans have nothing to do with plant respiration. Therefore, the correct answer is (C), "lenticels."

Although our example is a little goofy and doesn't look exactly like the questions you'll be seeing on the test, it illustrates an important point:

Process of Elimination is the best way to approach the multiple-choice questions.

This is true of all three portions of Section I. Even when you don't know the answer right off the bat, you'll surely know that two or three of the answer choices are not correct. What then?

STRATEGY 4: AGGRESSIVE GUESSING

ETS tells you that random guessing will not affect your score. This is true. In other words, if you guess on five problems, odds are you'll get one right. For the correct answer you'll receive one point, while for the four wrong answers, you'll lose one point (4 × [–1/4 point for each wrong answer] = –1 point). Net gain? Nothing!

However, the moment you've eliminated a couple of answer choices, your odds of getting the question right, even if you guess, are far greater. If you can eliminate as many as two answer choices, your odds improve enough that it's in your best interest to guess. How so? Let's look at an example.

Imagine that you've got three problems. On each problem, you've managed to eliminate two answer choices. If you guess on these three problems, you're bound to get one right, statistically speaking. For the correct answer you receive one point, while for the two wrong answers, you lose one-half a point (2 × [–1/4 point for each wrong answer] = –1/2 point). Net gain? One-half a point.

This may not seem like much, but if you do it aggressively throughout the Multiple-Choice section of the test, it could add as many as 10 to 15 points to your overall score. The difference between a decent test taker and an ace test taker is just this kind of aggressive approach.

STRATEGY 5: WORD ASSOCIATIONS

Another way to rack up the points on the AP Biology Exam is by using word associations in tandem with your POE skills. Make sure that you memorize the "Key Words" throughout this book. Know them backward and forward. As you learn them, make sure you group them by "association," since ETS is bound to ask about them on the AP Biology Exam. What do we mean by "word associations"? Let's take the example of mitosis and meiosis.

You'll soon see from our review that there are several terms associated with mitosis and meiosis. Synapsis, crossing-over, and tetrads, for example, are words associated with meiosis but not mitosis. We'll explain what these words mean in Chapter 7, in which we discuss reproduction. For now, just take a look:

2. Which of the following typifies cytokinesis during mitosis?

(A) Crossing-over
(B) Formation of the spindle
(C) Formation of tetrads
(D) Synapsis
(E) Division of the cytoplasm

This might seem like a difficult problem. But let's think about the associations we just discussed. The question asks us about mitosis. However, answer choices (A), (C), and (D) all mention events that we've associated with meiosis. Therefore, they are out. Without even racking your brain, you've managed to get this down to two answer choices. Not bad! For the record, the correct answer would then be (E), division of the cytoplasm.

Once again, don't worry about the science for now. We'll review it later. What is important to recognize is that by combining the associations we'll offer throughout this book and your aggressive POE techniques, you'll be able to rack up points on problems that might have seemed particularly difficult at first.

STRATEGY 6: MNEMONICS — OR THE BIOLOGY NAME GAME

One of the big keys to simplifying biology is the organization of terms into a handful of easily remembered packages. The best way to accomplish this is by using mnemonics. Biology is all about names: the names of chemical structures, processes, theories, etc. How are you going to keep them all straight? A mnemonic, as you may already know, is a convenient device for remembering something.

For example, one important issue in biology is "taxonomy," that is, the classification of life forms, or organisms. Organisms are classified in a descending system of similarity, leading from kingdoms (the broadest level) to species (the most specific level). The complete order runs: kingdom, phylum, class, order, family, genus, and species. Don't freak out yet. Look how easy it becomes with a mnemonic:

King Philip of Germany decided to walk to America. What do you think happened?

King	→	Kingdom
Philip	→	Phylum
Came	→	Class
Over	→	Order
From	→	Family
Germany	→	Genus
Soaked	→	Species

Learn the mnemonic and you'll never forget the science!

Mnemonics can be as goofy as you like, so long as they help you remember. Throughout this book, we'll give you mnemonics for many of the complicated terms we'll be seeing. Use ours, if you like them, or feel free to invent your own. Be creative! Remember: The important thing is that you remember the information, not how you remember it.

STRATEGY 7: IDENTIFYING QUESTION TYPES

Many of the traps on the AP Biology Exam deal with the way in which the question is asked.

EXCEPT/NOT/LEAST questions

About 10 percent of the multiple-choice questions in Section I are EXCEPT/NOT/LEAST questions. With this type of question, you must remember that: you're looking for the wrong (or the least correct) answer. The best way to go about these is by using POE.

More often than not, the correct answer is a true statement, but is wrong in the context of the question. However, the other four tend to be pretty straightforward. Cross off the four that apply and you're left with the one that does not. Here's a sample question:

17. All of the following are true statements about gametes EXCEPT:

(A) They are haploid cells.
(B) They are produced only in the reproductive structures.
(C) They bring about genetic variation among offspring.
(D) They develop from polar bodies.
(E) They combine to produce cells with the diploid number of chromosomes.

If you don't remember anything about gametes and gametogenesis, or the production of gametes, this might be a particularly difficult problem. We'll see these again later on, but for now, remember that gametes are the "sex cells" of sexually reproducing organisms. As such, we know that they are haploid and are produced in the sexual organs. We also know that they come together to create offspring.

From this very basic review, we know immediately that (A), (B), and (E) are not our answers. All three of these are accurate statements, so we eliminate them. That leaves us with (C) and (D). If you have no idea what (D) means, focus on (C). In sexual reproduction, each parent contributes one gamete, or half the genetic complement of the offspring. This definitely helps vary the genetic makeup of the offspring. Answer choice (C) is a true statement, and therefore wrong. The correct answer is (D).

Don't sweat it if you don't recall the biology. We'll be reviewing it in detail soon enough. For now, remember that the best way to answer these types of questions is: Spot all the right statements and cross them off. You'll wind up with the wrong statement, which happens to be the correct answer.

STRATEGY 8: THE ART OF THE ETS ESSAY

You're given four essay questions to answer in 90 minutes. As we saw above, that's only 22 minutes a question! The best way to rack up points on this section is to give the essay readers what they're looking for. Fortunately, we know precisely what that is.

The ETS essay reviewers have a checklist of key terms and concepts that they use to assign points. We like to call these "hot button" terms. Quite simply put, for each hot button that you include in your essay, you will receive a predetermined number of points. For example, if the essay question deals with the function of enzymes, the ETS graders are instructed to give 2 points for a mention of the "lock-and-key theory of enzyme specificity."

Naturally, you can't just compose a "laundry list" of scientific terms. Otherwise, it wouldn't be an essay. What you can do, however, is organize your essay around a handful of these key, or hot button, points. The most effective and efficient way to do this is by using the 10-minute reading period to brainstorm and come up with the scientific terms. Then outline your essay before you begin to write, using your hot buttons as your guide.

Brainstorm and outline

During the 10-minute preview, read each question twice and brainstorm the terms and concepts you want to cover with regard to each question. Once you've jotted down as many as you can, draft an outline that will help you organize them into some logical order.

Although the ETS graders do not grade you on your overall organization, a poorly organized essay tends to be less convincing than a well-organized one. The best way to avoid any problems with organization is to draw up a clear, simple outline. This can be done during the 10-minute reading period.

On average, you need to write one or two paragraphs for each part of the question. If the question asks for two examples, give just that—two examples. If you present more than two examples, the reviewers may not even count them toward your score. Make sure you read carefully and give them what they want!

Label diagrams and figures

Sometimes it's easier to present a diagram or figure as part of your essay. If you choose to do this, make sure you label your diagram or figure properly. Otherwise, the ETS graders will give you no more than partial credit for your work.

Review laboratory experiments covered in your AP course

At least one of the four questions will be experimentally based. Sometimes questions will refer back to a laboratory experiment conducted in your AP class. Consequently, the laboratory component of your course is an integral part of this exam. Don't forget that. In Chapter 15, we'll review some of the laboratory experiments (including equipment) that you're responsible for on the AP test.

Sample essays

Chapter 14 goes into far greater detail about the free-response questions. Take a look at it after you've reviewed the biology in the first part of this book.

There are sample essays as well as a checklist just like the one the ETS essay graders use when correcting your essay. Use them as guidelines when writing your own free-response essays for the practice test.

Correcting the essays isn't as clear-cut as correcting the multiple-choice questions. Nevertheless, by following our instructions, you'll be able to give yourself a rough idea of how you perform on this portion of the test.

LET'S GET CRACKING!

Along the way, we'll highlight what's important in each area, from cell structure to genetics to evolution. By helping you at each step, we take all the guesswork out of preparing for the test. In addition, as you'll soon see, we don't need to be dull and long-faced when it comes to biology. We can have some fun learning. Given the many advances that are being made in genetics, cell biology, and immunology, all this stuff is actually very interesting if it's looked at the right way. You may not believe it now, but by the end of this book, we're certain that you'll agree. (Especially when you see your test scores!)

We've done our work: We've taken the AP Biology Exam apart, pulled out the pieces you need to know, and presented them in an easy, accessible format. Now it's time for you to do your share. Follow along closely and answer all the questions at the end of each chapter. Answers to the quizzes and explanations for all the questions are found in Chapter 16 in the back of the book. If you learn the material on these pages, you're sure to improve your score.

Before we get started, let's do a quick summary of the strategies you need to remember for the test:

- **Pacing:** Know your pacing chart!

- **The Three-Pass System:** Focus your energy on the easy questions first—save the rest for later.

- **Process of Elimination:** Use POE to answer questions. Remember you don't need to know the right answers to get the questions right.

- **Aggressive Guessing:** Guess after you've eliminated two or more answer choices—it's in your best interest.

- **Word Associations:** Learn the lists at the end of each chapter. Know which words should be grouped together.

- **Mnemonics:** Use ours or make your own.

- **Identifying Question Types:** Look out for tricky questions, especially those EXCEPT/NOT/LEAST questions!

- **The Art of the ETS Essay:** Brainstorm, outline, write—and make sure you touch upon those "hot button" terms.

If you're comfortable with these strategies come test day, your score is bound to improve. Before we get there, however, we need to review the biology you'll see on the test. Without any further ado, let's get moving!

The Chemistry of Life

ELEMENTS

Although all of the substances in the universe are chemically diverse, they all have one thing in common: They're all made up of **elements**. Elements, by definition, are substances that cannot be broken down into simpler substances by chemical means. There are 92 naturally occurring elements found in nature.

THE ESSENTIAL ELEMENTS OF LIFE

Although there are 92 different elements in the known universe, 99 percent of all living matter is made up of just six of them. These are considered the essential elements of life:

Sulfur

Phosphorus

Oxygen

Nitrogen

Carbon

Hydrogen

…which gives us **SPONCH** (a nice, easy mnemonic!)

The elements listed above are ordered in the proportions in which they appear in living matter. That is, sulfur is the element least likely to be found in organisms, followed in increasing order by phosphorous, oxygen, etc., with hydrogen being the most abundant.

SUBATOMIC PARTICLES

If you break down an element into smaller pieces, you'll eventually come to the **atom**—the smallest unit of an element that retains its characteristic properties. Atoms are the building blocks of the physical world.

Within atoms, there are even smaller subatomic particles called **protons**, **neutrons**, and **electrons**. Let's take a look at a typical atom:

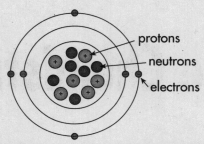

Protons and neutrons are particles that are packed together in the core of an atom called the **nucleus**. You'll notice that protons are positively charged (+) particles, whereas neutrons are uncharged particles.

Electrons, on the other hand, are negatively charged (–) particles that spin around the nucleus. Electrons are pretty small compared to protons and neutrons. In fact, for our purposes, electrons are considered massless. Most atoms have the same number of protons and electrons, making them electrically neutral.

COMPOUNDS

When two or more different types of atoms are combined in a fixed ratio, they form a chemical **compound**. You'll sometimes find that a compound has different properties from those of its elements. For instance, hydrogen and oxygen exist in nature as gases. Yet when they combine to make water, they often pass into a liquid state. When hydrogen atoms get together with oxygen atoms to form water, we've got a **chemical reaction**:

$$2H_2 \ (g) + O_2 \ (g) \rightarrow 2H_2O \ (l)$$

The atoms of a compound are held together by **chemical bonds**, which may be ionic bonds, covalent bonds, or hydrogen bonds.

An **ionic bond** is formed between two atoms when one or more electrons are transferred from one atom to the other. In this reaction, one atom *loses* electrons and becomes positively charged and the other atom *gains* electrons and becomes negatively charged. The charged forms of the atoms are called ions. For example, when Na reacts with Cl, charged ions, Na^+ and Cl^-, are formed.

A **covalent bond** is formed when electrons are *shared* between atoms. If the electrons are shared equally between the atoms the bond is called **nonpolar covalent**. If the electrons are shared unequally, the bond is called **polar covalent**. When one pair of electrons are shared, the result is a single bond. When two pairs of electrons are shared, the result is a double bond. When three pairs of electrons are shared, the result is a triple bond.

WATER: THE VERSATILE MOLECULE

One of the most important substances in nature is water. Did you know that 70 percent of your body weight consists of water? Water is considered a unique molecule because it plays an important role in chemical reactions.

Let's take a look at one of the properties of water. Water has two hydrogen atoms joined to an oxygen atom:

In water molecules, the hydrogen atoms have a partial positive charge and the oxygen atom has a partial negative charge. The positively-charged ends of the water molecule strongly attract the negatively-charged ends of other polar compounds. Likewise, the negatively-charged ends of water strongly attract the positively-charged ends of neighboring compounds. These forces are most readily apparent in the tendency of water molecules to stick together, as in the formation of water beads or raindrops.

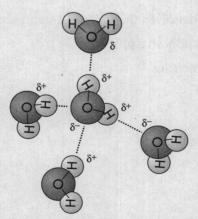

Molecules that have partially positive and partially negative charges are said to be **polar**. Water is therefore a polar molecule.

The polarity of water allows it to form weak bonds—called **hydrogen bonds**—with other polar substances. Hydrogen bonds are not actually bonds, but strong intermolecular forces that act in a bond-like way. It's this property that makes water a great solvent—it can dissolve many kinds of substances. The hydrogen bonds that hold water molecules contribute to a number of special properties:

- As mentioned above, water molecules have a strong tendency to stick together. That is, water exhibits *cohesive forces*. These forces are extremely important to life. For instance, when water molecules evaporate from a leaf, they "pull" neighboring water molecules. These, in turn, draw up the molecules immediately behind them, and so on, all the way down the plant vessels. The resulting chain of water molecules enables water to move up the stem.

- Water molecules also like to stick to other substances—that is, they're *adhesive*. Have you ever tried to separate two glass slides stuck together by a film of water? They're difficult to separate because of the water sticking to the glass surfaces. These two forces taken together—**cohesion** and **adhesion**—account for the ability of water to rise up the roots, trunks, and branches of trees. Since this phenomenon occurs in thin vessels, it's called **capillary action**.

- Another remarkable property of water is its high **heat capacity**. What's heat capacity? Your textbook will give you a definition something like this: "Heat capacity is the quantity of heat required to change the temperature of a substance by 1 degree." What does that mean? In plain English, heat capacity refers to the ability of a substance to store heat. For example, when you heat up an iron kettle, it gets hot pretty quickly. Why? Because it has a *low* specific heat. It doesn't take much heat to increase the temperature of the kettle. Water, on the other hand, has a high heat capacity. You have to add a lot of heat to get an increase in temperature. Water's ability to resist temperature changes is one of the things that helps keep the temperature in our oceans fairly stable. It's also why organisms that are mainly made up of water, like us, are able to keep a constant body temperature.

So let's review the unique properties of water:

- Water is polar and can dissolve other polar substances.
- Water has cohesive and adhesive properties.
- Water has a high heat capacity.

THE ACID TEST

We just said that water is important because most reactions occur in watery solutions. Well, there's one more thing to remember: Reactions are also influenced by whether the solution in which they occur is **acidic**, **basic**, or **neutral**.

What makes a solution acidic or basic? A solution is acidic if it contains a lot of *hydrogen ions* (H^+). That is, if you dissolve an acid in water, it will release a lot of hydrogen ions. When you think about acids, you usually think of substances with a sour taste, like lemons. For example, if you squeeze a little lemon juice into a glass of water, the solution will become acidic. That's because lemons contain citric acid.

Bases, on the other hand, do not release hydrogen ions when added to water. They release a lot of *hydroxide ions* (OH^-). These solutions are said to be **alkaline**. Bases usually have a slippery consistency. Common soap, for example, is composed largely of bases.

The acidity or alkalinity of a solution can be measured using a **pH scale**. The pH scale is numbered from 1 to 14. The midpoint, 7, is considered neutral pH. The concentration of hydrogen ions in a solution will indicate whether it is acidic, basic, or neutral. If a solution contains a lot of hydrogen ions, then it will be acidic and have a low pH. Here's the trend:

> An increase in H^+ ions causes a decrease in the pH.

You'll notice from the scale that stronger acids have lower pHs.

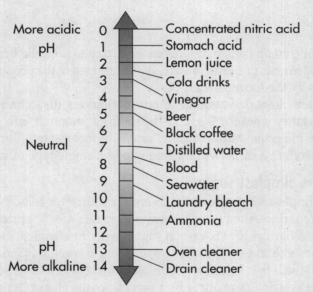

More acidic	0	Concentrated nitric acid
pH	1	Stomach acid
	2	Lemon juice
	3	Cola drinks
	4	Vinegar
	5	Beer
	6	Black coffee
Neutral	7	Distilled water
	8	Blood
	9	Seawater
	10	Laundry bleach
	11	Ammonia
	12	
pH	13	Oven cleaner
More alkaline	14	Drain cleaner

If a solution has a low concentration of hydrogen ions, it will have a high pH.

One more thing to remember: The pH scale is not a linear scale—it's logarithmic. That is, a change of *one* pH number actually represents a *tenfold* change in hydrogen ion concentration. For example, a pH of 3 is actually ten times *more* acidic than a pH of 4. This is also true in the reverse direction: A pH of 4 represents a tenfold *decrease* in acidity compared to a pH of 3.

ORGANIC MOLECULES

Now that we've discussed chemical compounds in general, let's talk about a special group of compounds. Most of the chemical compounds in living organisms contain skeletons of **carbon** atoms. These molecules are known as **organic compounds**. By contrast, molecules that do not contain carbon atoms are called **inorganic compounds**. For example, salt (NaCl) is an inorganic compound.

Carbon is important for life because it is a versatile atom, meaning that it has the ability to bind with other carbons as well as a number of other atoms. The resulting molecules are key in carrying out the activities necessary for life.

To recap:

- Organic compounds contain carbon atoms.

- Inorganic compounds do not contain carbon atoms (except carbon dioxide).

Now let's focus on four classes of organic compounds central to life on earth:

- Carbohydrates

- Amino acids

- Lipids

- Nucleic acids

Carbohydrates

Organic compounds that contain carbon, hydrogen, and oxygen are called carbohydrates. They usually contain these three elements in a ratio of 1:2:1. We can represent the proportion of elements within carbohydrate molecules by the formula $C_nH_{2n}O_n$.

Most carbohydrates are categorized as either **monosaccharides**, **disaccharides**, or **polysaccharides**. The term *saccharides* is a fancy word for "sugar." The prefixes "mono-," "di-," and "poly-" refer to the number of sugars in the molecule. Mono- means "one," di- means "two," and poly- means "many." A monosaccharide is therefore a carbohydrate made up of a single type of sugar molecule.

Monosaccharides: The simplest sugars

Monosaccharides, the simplest sugars, serve as an energy source for cells. While there are many different types of monosaccharides, the two most common sugars are (1) **glucose** and (2) **fructose**.

Both of these monosaccharides are six-carbon sugars with the chemical formula $C_6H_{12}O_6$. Glucose, the most abundant monosaccharide, is the most popular sugar around. Plants produce it by capturing sunlight for energy, while cells break it down to release stored energy. Glucose can come in two forms: α-glucose and β-glucose, which differ simply by a reversal of the H and OH on the first carbon. Fructose, the other monosaccharide you need to know for the test, is a common sugar in fruits.

One last thing. Glucose and fructose can be depicted as either "straight" or "rings." Both of them are pretty easy to spot; just look for the six carbon molecules. Here are the two different forms:

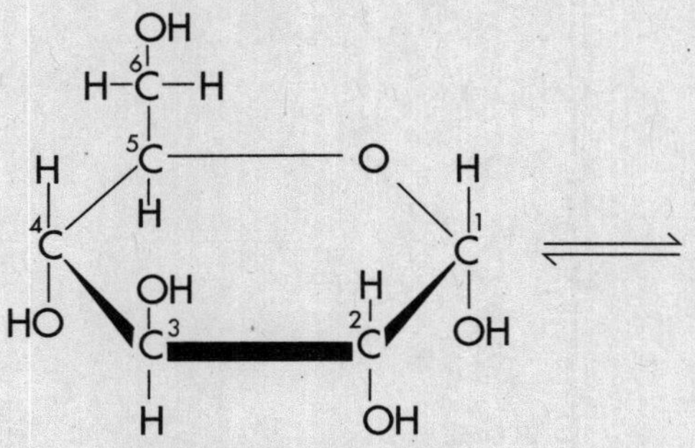

OH
H–C–H
6
5 C   O   H
H   H   1
4 C   C
HO   OH   H   OH
3 C —— 2 C
H   OH

Ring form of glucose

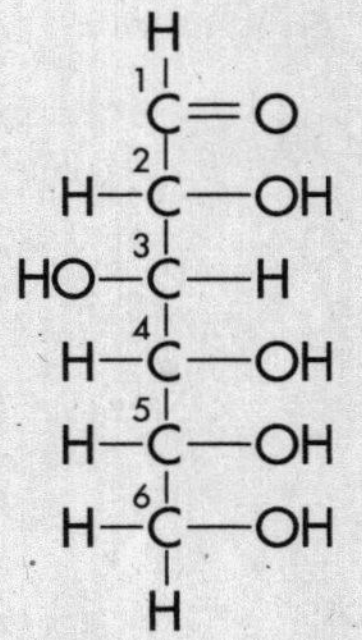

H
1 C=O
2
H–C–OH
3
HO–C–H
4
H–C–OH
5
H–C–OH
6
H–C–OH
H

Straight-chain
form of glucose

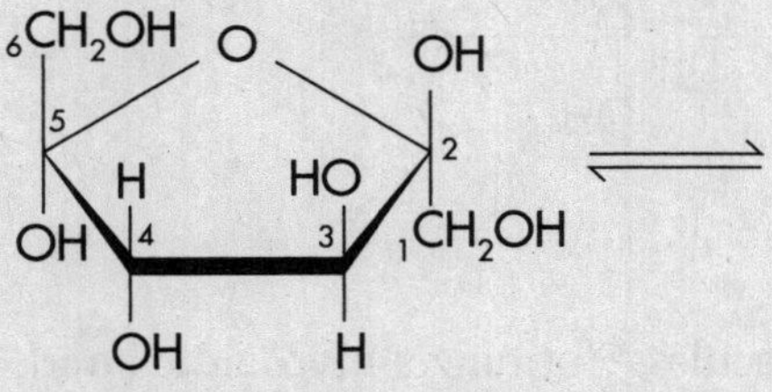

6 CH$_2$OH   O   OH
5   2
H   HO
OH 4   1 CH$_2$OH
3
OH   H

Ring form of fructose

H
1
H–C–OH
2
C=O
3
HO–C–H
4
H–C–OH
5
H–C–OH
6
H–C–OH
H

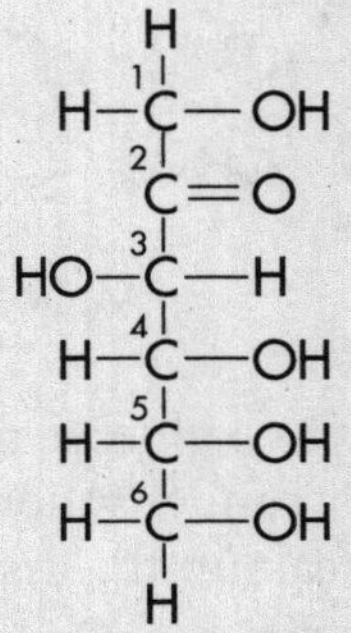

Straight-chain
form of fructose

Disaccharides

What happens when two monosaccharides are brought together? The hydrogen (–H) from one sugar molecule combines with the hydroxyl group (–OH) of another sugar molecule. What do H and OH add up to? Water (H_2O)! So a water molecule is removed from the two sugars. The two molecules of monosaccharides are chemically linked and form a disaccharide.

Maltose is an example of a disaccharide:

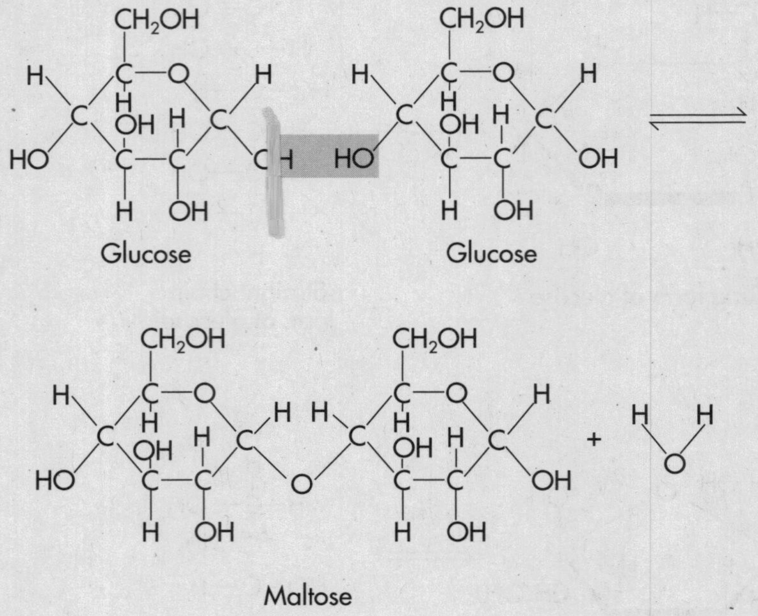

Glucose   Glucose

Maltose

Maltose is formed by linking two glucose molecules—forming a **glycosidic bond**. This process is called **dehydration synthesis**, or **condensation**.

Now what if you want to break up the disaccharide and form two monosaccharides again? Just add water. That's called **hydrolysis**.

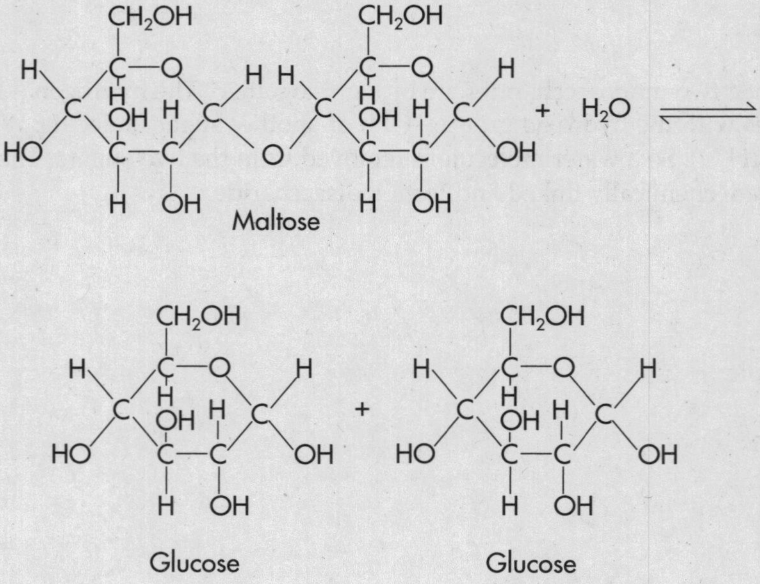

Maltose

Glucose   Glucose

Polysaccharides

Many biologically important carbohydrates called **polysaccharides** are made up of repeated units of simple sugars. The most common polysaccharides you'll need to know for the test are **starch, cellulose**, and **glycogen**. Polysaccharides are often storage forms of sugar or structural components of cells. For instance, animals store sugar in the form of glycogen in the liver and muscle cells. Plants "stockpile" α–glucose in the form of starch in structures called **plastids**. Cellulose, on the other hand, made up of β–glucose, is a major part of the cell wall in plants. Its function is to lend structural support.

Proteins

Amino acids are organic molecules that serve as the building blocks of proteins. They contain carbon, hydrogen, oxygen, and nitrogen atoms. There are 20 different amino acids commonly found in proteins. Fortunately, you don't have to memorize the 20 amino acids. But you do have to remember that every amino acid has four important parts: **an amino group** ($-NH_2$), **a carboxyl group** ($-COOH$), **a hydrogen**, and **an R group**.

Here's a typical amino acid:

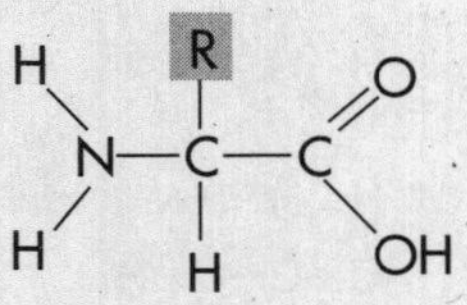

Amino acids differ only in the R group, which is also called the **side chain**. The R group associated with an amino acid could be as simple as a hydrogen atom (as in the amino acid *glycine*) or as complex as a carbon skeleton (as in the amino acid *arginine*).

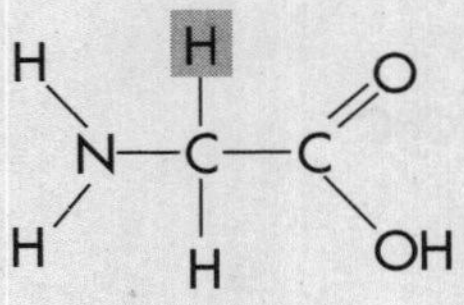

Glycine

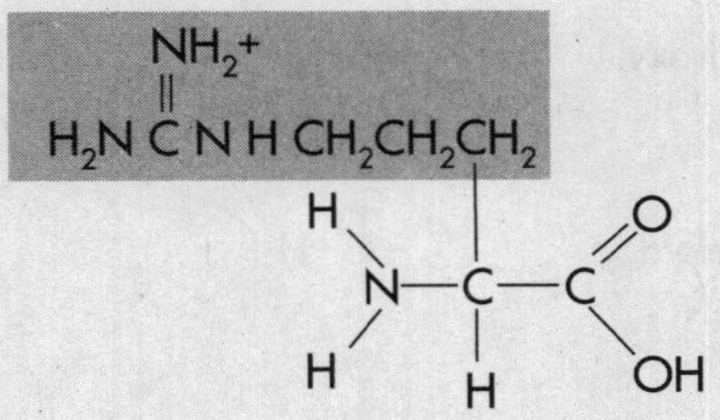

Arginine

Since ETS will probably test you on chemical diagrams, it's a good idea to flag the functional groups of the different structures. Functional groups are the distinctive groups of atoms that play a large role in determining the chemical behavior of the compound they are a part of. For example, an organic acid has a functional group, the carboxyl group (-COOH) that releases hydrogen ions in water. This makes the solution acidic. When it comes to spotting an amino acid, simply keep an eye out for the amino group (NH_2), then look for the carboxyl molecule (COOH). The most common functional groups of organic compounds are listed on the next page.

Name of Group	Structural Formula	Molecular Formula
Amino	$\begin{array}{c} H \\ | \\ -N-H \end{array}$	$-NH_2$
Alkyl	$\begin{array}{ccccc} H & H & & H & \\ | & | & & | & \\ -C-C- & \cdots & -C-H \\ | & | & & | & \\ H & H & & H & \end{array}$	$-C_nH_{2n+1}$
Methyl	$\begin{array}{c} H \\ | \\ -C-H \\ | \\ H \end{array}$	$-CH_3$
Ethyl	$\begin{array}{cc} H & H \\ | & | \\ -C-C-H \\ | & | \\ H & H \end{array}$	$-C_2H_5$
Propyl	$\begin{array}{ccc} H & H & H \\ | & | & | \\ -C-C-C-H \\ | & | & | \\ H & H & H \end{array}$	$-C_3H_7$
Carboxyl	$\begin{array}{c} O \\ \| \\ -C-O-H \end{array}$	$-COOH$
Hydroxyl	$-O-H$	$-OH$
Aldehyde	$\begin{array}{c} O \\ \| \\ -C-H \end{array}$	$-CHO$
Keto (carbonyl)	$\begin{array}{c} O \\ \| \\ -C- \end{array}$	$-CO$
Sulfhydryl (thiol)	$-S-H$	$-SH$
Phenyl	$\begin{array}{c} H \quad H \\ | \quad | \\ C-C \\ // \qquad \backslash \\ C \qquad\qquad C-H \\ \backslash \qquad // \\ C=C \\ | \quad | \\ H \quad H \end{array}$	$-C_6H_5$

Polypeptides

When two amino acids join they form a **dipeptide**. The carboxyl group of one amino acid combines with the amino group of another amino acid. Here's an example:

Here's the peptide bond

This is the same process we saw earlier: dehydration synthesis. Why? Because a water molecule is removed to form a bond. By the way, the bond between two amino acids has a special name—a **peptide bond**. If a group of amino acids are joined together in a "string," the resulting organic compound is called a **polypeptide**. Once a polypeptide chain twists and folds on itself, it forms a three-dimensional structure called a **protein**.

Lipids

Like carbohydrates, **lipids** consist of carbon, hydrogen, and oxygen atoms, but not in the 1:2:1 ratio typical of carbohydrates. The most common examples of lipids are **fats**, **oils**, **phospholipids**, and **steroids**. Let's talk about the simple lipids—**neutral fats**. A typical fat consists of three fatty acids and one molecule of **glycerol**. If you see the word "**triglyceride**" on the test, it's just a fancy word for "fat." Let's take a look:

To make a fat, each of the carboxyl groups (—COOH) of the three fatty acids must react with one of the three hydroxyl groups (—OH) of the glycerol molecule. This happens by the removal of a water molecule. Actually, a fat requires the removal of three molecules of water. Once again, what have we got? You probably already guessed it—dehydration synthesis! The bonds now formed between the glycerol molecule and the fatty acids are called **ester bonds**. A fatty acid can be **saturated**, which means it has a single covalent bond between each pair of carbon atoms or it can be **unsaturated**, which means adjacent carbon bonds are joined by double bonds instead of single bonds. A **polyunsaturated** fatty acid has many double bonds within the fatty acid.

Lipids are important because they function as structural components of cell membranes, sources of insulation, and a means of energy storage.

Phospholipids

Another special class of lipids is known as **phospholipids**. Phospholipids contain two fatty acid "tails" and one negatively charged phosphate "head." Take a look at a typical phospholipid:

glycerol   fatty acids

Phospholipids are extremely important, mainly because of some unique properties they possess, particularly with regard to water.

Interestingly enough, the two fatty acid "tails" are **hydrophobic** ("water-hating"). Why? Because fatty acids don't mix well with water. Just as oil and vinegar don't mix, neither do fatty acids and water. On the other hand, the phosphate "head" of the lipid is **hydrophilic** ("water-loving").

This arrangement of the fatty acid tails and the phosphate group head provides phospholipids with a unique shape. The two fatty acid chains orient themselves away from water, while the phosphate portion orients itself toward the water. Keep these properties in mind. We'll see later how this orientation of phospholipids in water relates to the structure and function of cell membranes.

One class of lipids is known as **steroids**. All steroids have a basic structure of four linked carbon rings. This category includes cholesterol, vitamin D, and a variety of hormones. Take a look at a typical steroid:

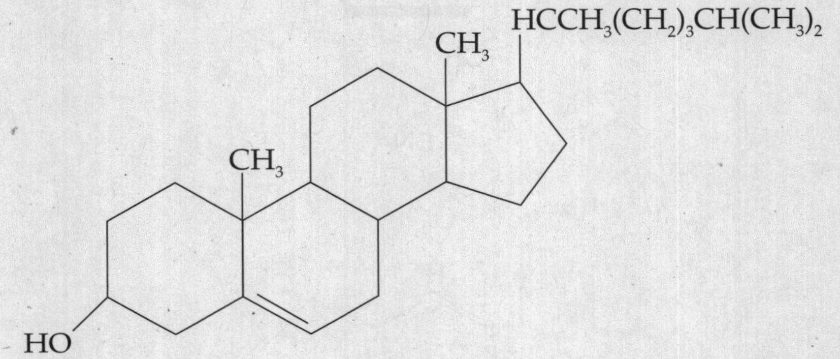

Nucleic acids

The fourth class of organic compounds are the **nucleic acids**. Like proteins, nucleic acids contain carbon, hydrogen, oxygen, and nitrogen, but nucleic acids also contain phosphorus. Nucleic acids are molecules that are made up of simple units called **nucleotides**. For the AP Biology Exam, you'll need to know about two kinds of nucleic acids: **deoxyribonucleic acid** (DNA) and **ribonucleic acid** (RNA).

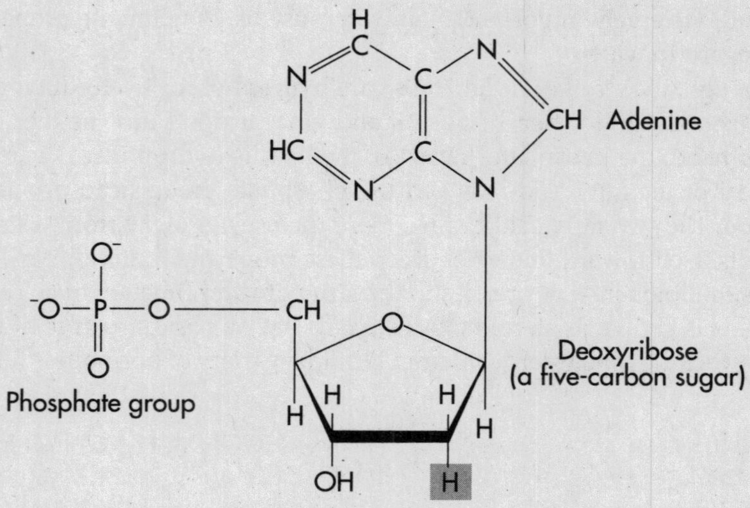

H

C

N   N

HC   CH Adenine

C   C

N   N

O⁻
|
⁻O—P—O—CH
|| |
O |

Phosphate group

O

H   H   H

H   H

OH   H

Deoxyribose
(a five-carbon sugar)

DNA

H

C

N   N

HC   CH Adenine

C   C

N   N

O⁻
|
⁻O—P—O—CH
|| |
O |

Phosphate group

O

H   H   H

H   H

OH   OH

Ribose
(a five-carbon sugar)

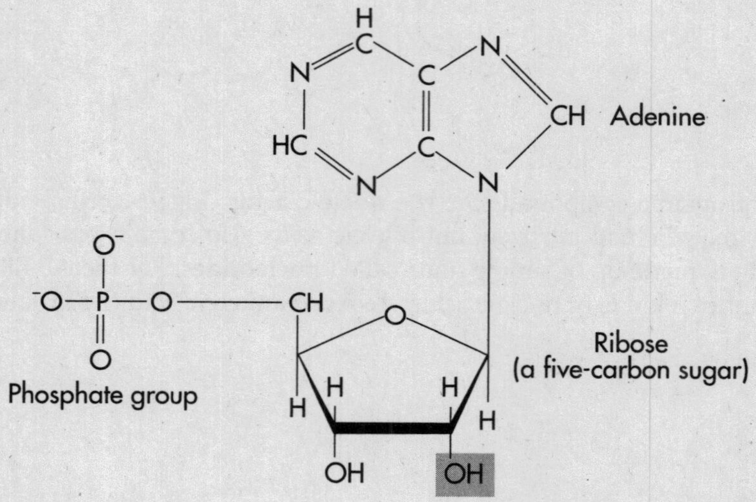

RNA

DNA is important because it contains genes, the hereditary "blueprints" of all life. RNA is important because it's essential for protein synthesis. We'll discuss DNA and RNA in greater detail when we discuss heredity.

THE HETEROTROPH HYPOTHESIS

We've just seen the organic compounds that are essential for life. But where did they come from in the first place? This is still a hotly debated topic among scientists. Most scientists believe that the earliest precursors of life arose from nonliving matter (basically gases) in the primitive oceans of the earth. But this theory didn't take shape until the 1920s. Two scientists, **Oparin** and **Haldane**, proposed that the primitive atmosphere contained the following gases: methane (CH_4), ammonia (NH_3), hydrogen (H_2), and water (H_2O). Interestingly enough, there was almost no free oxygen (O_2) in this early atmosphere. They believed that these gases collided, producing chemical reactions that eventually led to the organic molecules we know today.

This theory didn't receive any substantial support until 1953. In that year, **Stanley Miller** and **Harold Urey** simulated the conditions of primitive earth in a laboratory. They put the gases theorized to be abundant in the early atmosphere into a flask, struck them with electrical charges in order to mimic lightning, and organic compounds similar to amino acids appeared!

But how do we make the leap from simple organic molecules to more complex compounds and life as we know it? Since no one was around to witness the process, no one knows for sure how (or when) it occurred. Complex organic compounds (such as proteins) must have formed via dehydration synthesis. Simple cells then used organic molecules as their source of food. Over time, simple cells evolved into complex cells.

Now let's throw in a few new terms. Living organisms that rely on organic molecules for food are called **heterotrophs.** For example, we're heterotrophs. Eventually, some organisms (such as plants) found a way to make their own food. These organisms are called **autotrophs** (or producers).

Although the precise origins of life may remain forever unexplained, scientists believe that the earliest life forms were most likely heterotrophs, relying on other organic molecules for energy. This theory is known the **heterotroph hypothesis**.

KEY WORDS

elements
atom
protons
neutrons
nucleus
electrons
compound
chemical reaction
polar
hydrogen bonds
cohesion
adhesion
capillary action
heat capacity
acidic
basic
neutral
alkaline
pH scale

organic compounds
inorganic compounds
carbohydrates
monosaccharides
disaccharides
polysaccharides
glucose
fructose
glycosidic bond
dehydration synthesis
   (or condensation)
hydrolysis
starch
cellulose
glycogen
plastids
amino acids
functional groups
dipeptide

peptide bond
polypeptide
protein
lipids
fat
ester bonds
phospholipids
hydrophobic
hydrophilic
nucleic acids
nucleotides
DNA (deoxyribonucleic acid)
RNA (ribonucleic acid)
Oparin and Haldane
Stanley Miller
Harold Urey
heterotrophs
autotrophs (or producers)
heterotroph hypothesis

QUIZ

<u>Directions:</u> Each group of questions consists of five lettered headings followed by a list of numbered phrases or sentences. For each numbered phrase or sentence, select the one heading that is most closely related to it and fill in the corresponding oval on the answer sheet. Each heading may be used once, more than once, or not at all in each group.

Questions 1–4

    (A) Hydrogen bond
    (B) Peptide bond
    (C) Glycosidic bond
    (D) Ester bond
    (E) Amino group

1. Weak bond formed between polar molecules

2. Bond linking two monosaccharides in glycogen

3. The connecting –CO–NH bond in an organic molecule

4. Bond that links a fatty acid to a glycerol molecule

5. Basic functional group of organic molecules

<u>Directions:</u> Each of the questions or incomplete statements below is followed by five suggested answers or completions. Select the one that is best in each case. Answers can be found on page 236.

6. Which of the following organic molecules is a major storage carbohydrate used to store energy in plants?

    (A) Cellulose
    (B) Maltose
    (C) Fructose
    (D) Starch
    (E) Glycogen

7. A solution with a pH of 10 is how many times more basic than a solution with a pH of 8?

    (A)   2
    (B)   4
    (C)   10
    (D)   100
    (E)   1000

8. The conversion of lactose (a disaccharide) to glucose and galactose (monosaccharides) involves the addition of which of the following species to the lactose molecule?

(A) O_2
(B) H_2
(C) ATP
(D) H_2O
(E) NADH

9. All of the following organic compounds are polymers EXCEPT:

(A) starch
(B) cellulose
(C) polypeptide
(D) glycine
(E) glycogen

10. All of the following organic compounds contain a hydroxyl functional group EXCEPT:

(A) maltose
(B) glucose
(C) fructose
(D) glycerol
(E) triglyceride

11. A particular polypeptide contains 90 amino acids. When the polypeptide is completely hydrolyzed, how many water molecules are formed during this process?

(A) 2
(B) 30
(C) 45
(D) 89
(E) 90

12. All of the following qualities contribute to the unique properties of water EXCEPT:

(A) cohesion
(B) adhesion
(C) polarity
(D) capillary action
(E) low heat capacity

3

Cells and Organelles

LIVING THINGS

All living things—plants and animals—are composed of **cells**. According to the cell theory, the cell is life's basic unit of structure and function. This simply means that the cell is the smallest unit of living material that can carry out all the activities necessary for life.

WHAT ARE THE DIFFERENT TYPES OF CELLS?

For centuries, scientists have known about cells. However, it wasn't until the development of the electron microscope that scientists were able to figure out what cells do. We now know that there are two distinct types of cells: **eukaryotic cells** and **prokaryotic cells**. A eukaryotic cell contains a membrane-bound structure called a **nucleus** and **cytoplasm**, filled with tiny structures called **organelles** (literally "little organs"). Examples of eukaryotic cells are *fungi, protists, plant cells*, and *animal cells*.

A prokaryotic cell, which is a lot smaller than a eukaryotic cell, lacks both a nucleus and membrane-bound organelles. The genetic material in a prokaryote is one continuous, circular DNA molecule that lies free in the cell in an area called the **nucleoid**. In addition to a plasma membrane, most prokaryotes have a cell wall composed of peptidoglycan. Prokaryotes may also have ribosomes (although smaller than those found in eukaryotic cells) as well as a flagellum, a long fiber that helps them move.

ORGANELLES

A eukaryotic cell is sort of like a microscopic factory. It's filled with organelles, each of which has its own special tasks. Let's take a tour of a eukaryotic cell and focus on the structure and function of each organelle. Here's a picture of a typical cell and its principal organelles:

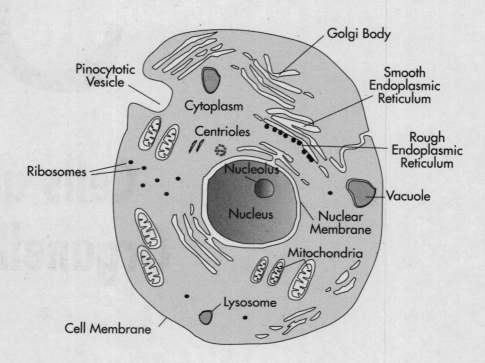

Plasma membrane

The cell has an outer envelope known as the **plasma membrane**. Although the plasma membrane appears to be a simple, thin layer surrounding the cell, it's actually a complex double-layered structure made up of phospholipids and proteins. The *hydrophobic* fatty acid tails face inward and the *hydrophilic* phosphate heads face outward.

Throughout the plasma membrane are various proteins that float through the phospholipid matrix. Some of these proteins are loosely associated with the lipid bilayer (**peripheral proteins**). They are located on the inner or outer surface of the membrane. Others are firmly bound to the plasma membrane (**integral proteins**). These proteins are amphipathic, which means that their hydrophilic regions extend out of the cell or into the cyptoplasm while their hydrophobic regions interact with the tails of the membrane phospholipids. Some integral proteins do not extend all the way through the membrane (**transmembrane proteins**). This arrangement of phospholipids and proteins is known as the **fluid-mosaic model**.

Why should the plasma membrane need so many different proteins? It's because of the number of activities that take place in or on the membrane. Generally, plasma membrane proteins fall into several broad functional groups. Some membrane proteins form junctions between adjacent cells (**adhesion proteins**). Others serve as docking sites for proteins of the extracellular matrix or hormones (**receptor proteins**). Some proteins form pumps that use ATP to actively transport solutes across the membrane (**transport proteins**). Others form channels that selectively allow the passage of certain ions or molecules (**channel proteins**). Finally, some proteins, such as glycoproteins, are exposed on the extracellular surface and play a role in cell recognition and adhesion (**recognition and adhesion proteins**).

Cholesterol molecules are also found in the phospholipid bilayer because they help stabilize membrane fluidity in animal cells.

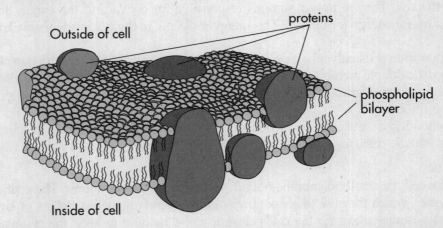

The plasma membrane is important because it regulates the movement of substances into and out of the cell. The membrane itself is semi-permeable, meaning that only certain substances pass through it unaided.

The nucleus

The **nucleus**, which is usually the largest organelle, is the control center of the cell. The nucleus not only directs what goes on in the cell, it is also responsible for the cell's ability to reproduce. It's the home of the hereditary information—DNA—which is organized into large structures called **chromosomes**. The most visible structure within the nucleus is the **nucleolus**, which is where rRNA is made and ribosomes are assembled.

Ribosomes

The **ribosomes** are the sites of protein synthesis. Their job is to manufacture all the proteins required by the cell or secreted by the cell. Ribosomes are round structures composed of RNA and proteins. They can be either free floating in the cell or attached to another structure called the **endoplasmic reticulum** (ER).

Endoplasmic reticulum (ER)

Not all proteins are synthesized on freely floating ribosomes. In fact, some are synthesized by ribosomes attached to the surface of the endoplasmic reticulum (ER). The ER is a continuous channel that extends into many regions of the cytoplasm. If the ER is "studded" with ribosomes, it's called the rough ER (RER). Proteins made on the *rough ER* are the ones "earmarked" to be exported out of the cell. If the ER lacks ribosomes, it's called the *smooth ER* (SER). The smooth ER makes lipids, hormones, steroids and breaks down toxic chemicals.

Golgi bodies

The **Golgi bodies**, which look like stacks of flattened sacs, also participate in the processing of proteins. Once the ribosomes on the rough ER have completed synthesizing proteins, the Golgi bodies modify, process, and sort the products. They're the packaging and distribution centers for materials destined to be sent out of the cell. They package the final products in little sacs called **vesicles**, which carry the products to the plasma membrane.

Mitochondria: The powerhouses of the cell

Another important organelle is the *mitochondrion*. The **mitochondria** are often referred to as the "powerhouses" of the cell. They're power stations responsible for converting the energy from organic molecules into useful energy for the cell. The energy molecule in the cell is **adenosine triphosphate, (ATP)**.

The mitochondrion is usually an easy organelle to recognize because it has a unique oblong shape and a characteristic double membrane consisting of an inner portion and an outer portion. The inner mitochondrial membrane forms folds known as *cristae*. As we'll see later, most of the production of ATP is done on the cristae.

Since mitochondria are the cell's powerhouses, you're most likely to find them in cells that require a lot of energy. Muscle cells, for example, are rich in mitochondria.

Lysosomes

Throughout the cell are small, membrane-bound structures called **lysosomes**. These tiny sacs carry digestive enzymes, which they use to break down old, worn-out organelles, debris, or large ingested particles. The lysosomes make up the cell's cleanup crew, helping to keep the cytoplasm clear of unwanted flotsam.

Centrioles

The **centrioles** are small, paired cylindrical structures that are found within **microtubule organizing centers (MTOCs)**. Centrioles are most active during cellular division. When a cell is ready to divide, the centrioles produce *microtubules*, which pull the replicated chromosomes apart and move them to opposite ends of the cell. Although centrioles are common in animal cells, they are not found in plant cells.

Vacuoles

In Latin, the term vacuole means "empty cavity." But **vacuoles** are far from empty. They are fluid filled sacs that store water, food, wastes, salts, or pigments.

Peroxisomes

Peroxisomes are organelles that detoxify various substances, producing hydrogen peroxide as a byproduct. They also contain enzymes that break down hydrogen peroxide (H_2O_2) into oxygen and water. In animals, they are common in the liver and kidney cells.

Cytoskeleton

Have you ever wondered what actually holds the cell together and enables it to keep its shape? The shape of a cell is determined by a network of fibers called the **cytoskeleton**. The most important fibers you'll need to know are **microtubules** and **microfilaments**.

Microtubules, which are made up of the protein **tubulin**, participate in cellular division and movement. These small fibers are an integral part of three structures: *centrioles*, *cilia*, and *flagella*. We've already mentioned that centrioles help chromosomes separate during cell division. Cilia and flagella, on the other hand, are threadlike structures best known for their locomotive properties in single-celled organisms. The beating motion of cilia and flagella structures propels these organisms through their watery environments.

The two classic examples of organisms with these structures are the *euglena*, which gets about using its whip-like flagellum, and the *paramecium*, which is covered in cilia. The rhythmic beating of the paramecium's cilia enables it to motor about in waterways, ponds, and microscope slides in your biology lab. You've probably already checked these out in lab, but here's what they look like:

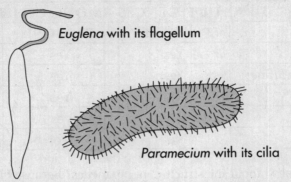

Euglena with its flagellum

Paramecium with its cilia

Though we usually associate such structures with microscopic organisms, they aren't the only ones with cilia and flagella. As you probably know, these structures are also found in certain human cells. For example, the cells lining your respiratory tract possess cilia that sweep constantly back and forth (beating up to 20 times a second), helping to keep dust and unwanted debris from descending into your lungs. And every sperm cell has a flagellum, which enables it to swim through the female reproductive organs to fertilize the waiting ovum.

Microfilaments, like microtubules, are important for movement. These thin, rodlike structures are composed of the protein actin, they are involved in cell mobility, and play a central role in muscle contraction.

PLANT CELLS VERSUS ANIMAL CELLS

Plant cells contain most of the same organelles and structures seen in animal cells, with several key exceptions. Plant cells, unlike animal cells, have a protective outer covering called the **cell wall** (made of cellulose). A cell wall is a rigid layer just outside of the plasma membrane that provides support for the cell. It is found in plants, protists, fungi, and bacteria. (In fungi, the cell wall is usually made of **chitin**, a modified polysaccharide. Chitin is also a principle component of an arthropod's exoskeleton.) In addition, plant cells possess **chloroplasts** (organelles involved in photosynthesis). Chloroplasts contain chlorophyll, the light-capturing pigment that gives plants their characteristic green color. Another difference between plant and animal cells is that most of the cytoplasm within a plant cell is usually taken up by a large vacuole that crowds the other organelles. In mature plants, this vacuole contains the **cell sap**. Plant cells also differ from animal cells in that plant cells do not contain centrioles.

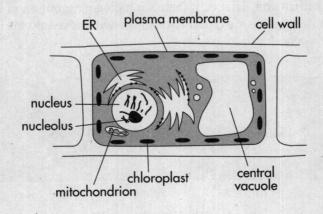

ER   plasma membrane   cell wall

nucleus

nucleolus

mitochondrion   chloroplast   central vacuole

To help you remember the differences among prokaryotes, plant cells, and animal cells, we've put together this simple table. Make sure you learn it! ETS is bound to ask you which cells contain which structures:

STRUCTURAL DIFFERENCES AMONG CELL TYPES			
Structure	Prokaryote	Plant Cell	Animal Cell
Cell Wall	Yes	Yes	No!
Plasma Membrane	Yes	Yes	Yes
Organelles	No!	Yes	Yes
Nucleus	No!	yes	Yes
Centrioles	No!	No!	Yes

Why do we need to know about the structure of organelles? Because biological structure is often closely related to function. (Watch out for this connection: It's a favorite theme for the AP Biology Exam.) And, more important, because ETS likes to test you on it!

TRANSPORT: THE TRAFFIC ACROSS MEMBRANES

We've talked about the structure of cell membranes, now let's discuss how molecules and fluids pass through the plasma membrane. What are some of the patterns of membrane transport? The ability of molecules to move across the cell membrane depends on two things: (1) the permeability of the plasma membrane and (2) the size and charge of particles that want to get through.

First let's consider how cell membranes work. For a cell to maintain its internal environment, it has to be selective in the materials it allows to cross its membrane. Since the plasma membrane is composed primarily of phospholipids, lipid-soluble substances cross the membrane without any resistance. Why? Because "like dissolves like." Generally speaking, the lipid membrane has an open-door policy for substances that are made up of lipids. These substances can cross the plasma membrane without any problem. However, if a substance is not lipid-soluble, the bilipid layer won't let it in.

One exception to the rule is water:

> Although water molecules are polar (and therefore not lipid-soluble) they can rapidly cross a lipid bilayer because they are small enough to pass through gaps that occur as a fatty acid chain momentarily moves out of the way.

Diffusion

We've just seen that lipid-soluble substances can traverse the plasma membrane without much difficulty. But what determines the *direction* of traffic across the membrane? Some substances move across a membrane by **simple diffusion**. That is, if there's a high concentration of a substance outside the cell and a low concentration inside the cell, the substance will move into the cell. In other words, the substance moves *down* a concentration gradient:

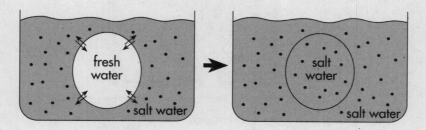

It's like riding a bicycle downhill. The bike "goes with the natural flow." Another name for this type of transport is **passive transport**. Here's one more thing you must remember:

Simple diffusion does not require energy.

A special kind of diffusion that involves the movement of a liquid (such as water) is called **osmosis**.

Facilitated transport

How do substances that are not lipid-soluble cross the plasma membrane? They've come up with another way to get in. Remember the proteins embedded in the plasma membrane? Well, special proteins—called **channel proteins**—can help lipid-insoluble substances or "solutes" get across the membrane:

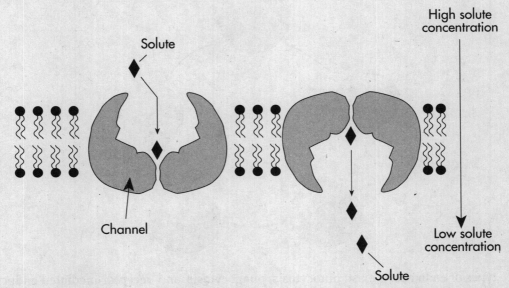

These proteins pick up the substance from one side of the membrane and carry it across to the other. This type of transport is known as **facilitated transport**, or facilitated diffusion. Facilitated diffusion is just like simple diffusion in one respect: The flow of the substance is *down* the concentration gradient. Therefore, it doesn't require any energy.

Active transport

Suppose a substance wants to move in the opposite direction—from a region of lower concentration to a region of higher concentration. It's going to need energy to accomplish this. This time it's like riding a bicycle uphill. Compared with riding downhill, riding uphill takes a lot more work. Movement against the natural flow is called **active transport**.

But where does the cell get this energy? Some proteins in the plasma membrane are powered by ATP. The best example of active transport is a special protein called the **sodium-potassium pump**. It ushers out sodium ions (Na^+) and brings in potassium ions (K^+) across the cell membrane. These pumps *depend* on ATP to get ions across that would otherwise remain in regions of higher concentration. Where do we usually find these pumps?

We've now seen that small substances can cross the cell membrane by:

- Simple diffusion
- Facilitated transport
- Active transport

Endocytosis

When the particles that want to enter a cell are just too large, the cell uses a portion of the cell membrane to engulf the substance. The cell membrane forms a pocket, pinches in, and eventually forms either a vacuole or a vesicle. This is called **endocytosis**.

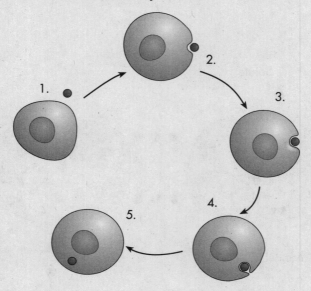

Three types of endocytosis exist: **pinocytosis**, **phagocytosis**, and **receptor-mediated endocytosis**. In pinocytosis, the cell ingests liquids ("cell-drinking"). In phagocytosis, the cell takes in solids ("cell-eating"). A special type of endocytosis, receptor-mediated endocytosis, begins at a receptor when a particle, called a ligand, binds to the cell's surface receptor that contains clathrin-coated pits. The ligand is then brought into the cell by the invagination of the cell membrane. A vesicle then forms around the incoming ligand and carries it into the cell's interior.

Bulk flow

Other substances move by bulk flow. Bulk flow is the one-way movement of fluids brought about by *pressure*. For instance, the movement of blood through a blood vessel or movement of fluids in xylem and phloem of plants are examples of bulk flow.

Dialysis

Dialysis is the diffusion of *solutes* across a selectively permeable membrane. For example, a cellophane bag is often used as an artificial membrane to separate small molecules from large molecules.

Exocytosis

Sometimes large particles are transported *out* of the cell. In **exocytosis**, a cell ejects waste products or specific secretion products such as hormones by the fusion of a vesicle with the plasma membrane.

CELL JUNCTIONS

When cells come in close contact with each other, they develop specialized **intercellular junctions** that involve their plasma membranes as well as other components. These structures may allow neighboring cells to form strong connections with each other, prevent passage of materials, or establish rapid communication between adjacent cells. There are three types of intercellular contact in animal cells: **desmosomes**, **gap junctions**, and **tight junctions**.

Desmosomes hold adjacent animal cells tightly to each other, like a rivet. They consist of a pair of discs associated with the plasma membrane of adjacent cells, plus the intercellular protein filaments that cross the small space between them. Intermediate filaments within the cells are also attached to the discs (Figure 1).

Gap junctions are protein complexes that form channels in membranes and allow communication between the cytoplasm of adjacent animal cells or the transfer of small molecules and ions.

Tight junctions are tight connections between the membranes of adjacent animal cells. They're so tight that there is no space between the cells. Cells connected by tight junctions seal off body cavities and prevent leaks.

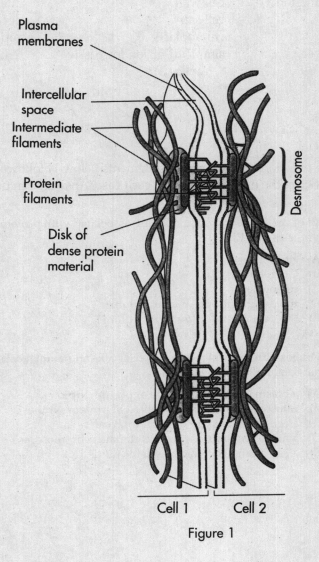

Plasma membranes

Intercellular space

Intermediate filaments

Protein filaments

Disk of dense protein material

Desmosome

Cell 1   Cell 2

Figure 1

KEY WORDS

cells
eukaryotic cells
prokaryotic cells
nucleus
cytoplasm
organelles
plasma membrane
peripheral proteins
integral proteins
transmembrane proteins
adhesion proteins
receptor proteins
channel proteins
recognition and adhesion
  proteins
cholesterol
phospholipid bilayer

fluid-mosaic model
chromosomes
nucleolus
ribosomes
endoplasmic reticulum (ER)
Golgi bodies
vesicles
mitochondria
adenosine triphosphate (ATP)
lysosomes
centrioles
vacuoles
cytoskeleton
cell wall
cell sap
chloroplasts
simple diffusion (or passive
  transport)

osmosis
channel proteins
facilitated transport (or facilitated diffusion)
active transport
sodium-potassium pump
endocytosis
pinocytosis
phagocytosis
receptor-mediated endocytosis
bulk flow
dialysis
exocytosis
intercellular junctions
desmosomes
gap junctions
tight junctions

QUIZ

<u>Directions:</u> Each of the questions or incomplete statements below is followed by five suggested answers or completions. Select the one that is best in each case. Answers can be found on page 237.

1. All of the following organelles are associated with protein synthesis EXCEPT:

    (A) ribosomes
    (B) Golgi bodies
    (C) the nucleus
    (D) the rough endoplasmic reticulum
    (E) the smooth endoplasmic reticulum

2. A major difference between bacterial cells and animal cells is that bacterial cells have:

    (A) a plasma membrane made of phospholipids
    (B) ribosomes, which are involved in protein synthesis
    (C) a cell wall made up of peptidoglycan
    (D) a nuclear membrane, which contains chromosomes
    (E) a large vacuole, which contains fluids

3. All of the following are methods of moving materials across cell membranes without the expenditure of energy EXCEPT:

(A) exocytosis
(B) facilitated transport
(C) osmosis
(D) active transport
(E) diffusion

Directions: Each group of questions consists of five lettered headings followed by a list of numbered phrases or sentences. For each numbered phrase or sentence, select the one heading that is most closely related to it and fill in the corresponding oval on the answer sheet. Each heading may be used once, more than once, or not at all in each group.

Questions 4–8

(A) Smooth ER
(B) Lysosome
(C) Cell wall
(D) Microtubule
(E) Nucleolus

4. Vesicles that serve to break down cellular debris

5. A semi-rigid structure that lends support to a cell

6. A channel inside the cytoplasm that is the site of lipid synthesis

7. A polymer of the protein tubulin that is found in cilia, flagella, and spindle fibers

8. Site at which rRNA is formed

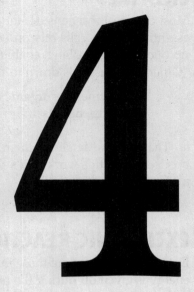

Cellular Metabolism

BIOENERGETICS

In Chapter 2, we discussed some of the more important organic molecules. But what makes these molecules so important? Glucose, starch, and fat are all energy-rich. However, the energy is packed in the chemical bonds holding the molecules together. To carry out the processes necessary for life, cells must find a way to release the energy in these bonds when they need it and store it away when they don't. The study of how cells accomplish this is called **bioenergetics**. In its most general sense:

> Bioenergetics is the study of how energy from the sun is transformed into energy in living things.

All the energy for life comes from chemical bonds. During chemical reactions, such bonds are either broken or formed. This process involves energy, no matter which direction we go in:

> Every chemical reaction involves a change in energy.

Energy is invested in the formation of bonds, whereas energy is released when bonds are broken. But even the breaking apart of chemical bonds requires the input of some energy. How, then, do cells manage to get the energy they need for life? They get it, in large part, from proteins called **enzymes**.

ENZYMES

Most chemical reactions do not occur haphazardly in the cell. To help control the chemical reactions essential for life, cells rely on enzymes. These proteins help "kick-start" reactions and speed them up once they get rolling, enabling cells to get the most out of the energy sources available to them. "Officially," this boils down to the following definition:

> Enzymes are **organic catalysts**; they speed up the rate of a reaction without altering the reaction itself.

That is, they "catalyze" them without being changed in the reaction themselves. Before we discuss precisely how this works, let's review some of the different types of reactions that can occur within a cell.

EXERGONIC REACTIONS

Exergonic reactions are those in which the products have *less* energy than the reactants. Simply put, energy is given off during the reaction.

Let's look at an example. The course of a reaction can be represented by an energy diagram. Here's an energy diagram for an exergonic reaction:

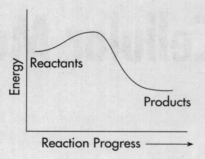

You'll notice that energy is represented along the *y*-axis. Based on the diagram, our reaction released energy. An example of an exergonic reaction is when food is oxidized in mitochondria of cells and then releases the energy stored in the chemical bonds.

Reactions that require an input of energy are called **endergonic reactions**. You'll notice that the *products* have more energy than the *reactants*.

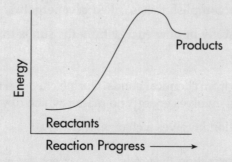

The products gained energy in the form of heat. An example is when plants use carbon dioxide and water to form sugars.

ACTIVATION ENERGY

Although we said that exergonic reactions release energy, this does not mean that they do not require any energy to get started. Take a look at this energy diagram of a typical exergonic reaction:

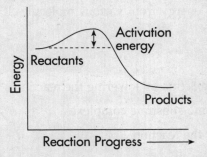

You'll notice that we needed a little energy to get us going. That's because chemical bonds must be broken before new bonds can form. This energy barrier—the hump in the graph—is called the **activation energy**. Once a set of reactants has reached its activation energy, the reaction can occur much faster than it would in the absence of the enzyme.

GETTING BACK TO ENZYMES...

Why are enzymes so important in biology? They're important because many reactions would never occur in the cell if it weren't for the help of enzymes. As we saw earlier, enzymes, by definition, catalyze reactions: They "kick-start," or activate, them. What this means in chemical terms is that they *lower* the activation energy of a reaction, enabling the reaction to occur much faster than it would in the absence of the enzyme.

ENZYME SPECIFICITY

Most of the crucial reactions that occur in the cell require enzymes. Yet enzymes themselves are highly specific—in fact, each enzyme catalyzes only one reaction. This is known as **enzyme specificity**. Since this is true, enzymes are usually named after the molecules they target. In enzymatic reactions, the targeted molecules are known as **substrates**. For example, maltose, a disaccharide, can be broken down into two glucose molecules. Our substrate, maltose, gives its name to the enzyme that catalyzes this reaction: *maltase*.

Many enzymes are named simply by replacing the suffix of the substrate with "*-ase*." Using this nomenclature, malt*ose* becomes malt*ase*.

ENZYME-SUBSTRATE COMPLEX

Enzymes have a unique way of helping reactions along. As we just saw, the reactants in an enzyme-assisted reaction are known as *substrates*. During a reaction, the enzyme's job is to bring the substrates together. It accomplishes this due to a special region on the enzyme known as an **active site**.

The enzyme temporarily binds the substrates to its active site, and forms an **enzyme-substrate complex**. Let's take a look:

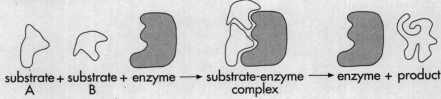

Once the reaction has occurred and the product is formed, the enzyme is released from the complex and restored to its original state. Now, the enzyme is free to react again with another bunch of substrates.

By binding and releasing over and over again, the enzyme speeds the reaction along, enabling the cell to release much-needed energy from various molecules. How about a quick review on the function of enzymes:

Enzymes Do:

- Increase the rate of a reaction by lowering the reaction's activation energy
- Form temporary enzyme-substrate complexes
- Remain unaffected by the reaction

Enzymes Don't:

- Change the reaction
- Make reactions occur that would otherwise not occur at all

INDUCED FIT

However, scientists have discovered that enzymes and substrates don't fit together quite so seamlessly. It appears that the enzyme has to change its shape slightly to accommodate the shape of the substrates. This is called **induced fit**.

ENZYMES DON'T ALWAYS WORK ALONE

Enzymes sometimes need a little help in catalyzing a reaction. Those factors are known as **coenzymes**. Vitamins are examples of organic coenzymes. Your daily dose of vitamins is important for just this reason: The vitamins are "active and necessary participants" in crucial chemical reactions. The function of coenzymes is to accept electrons and pass them along to another substrate. Two examples of such enzymes are NAD^+ and $NADP^+$.

In addition to organic coenzymes, inorganic elements—called **cofactors**—help catalyze reactions. These elements are usually metal ions, such as Fe^{+2}.

FACTORS AFFECTING REACTION RATES

Enzymatic reactions can be influenced by a number of factors, such as temperature, pH, and the relative amounts of enzyme and substrate.

Temperature

The rate of a reaction increases with increasing temperature because an increase in the temperature of a reaction increases the chance of collisions among the molecules. Here's one thing to remember: All enzymes operate at an ideal temperature. For most human enzymes, this temperature is body temperature, 37°C.

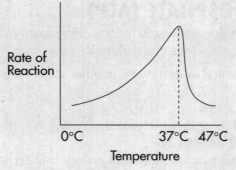

But too much heat can damage an enzyme. If a reaction is conducted at an excessively high temperature (above 42°C), the enzyme loses its three-dimensional shape and becomes inactivated. Enzymes damaged by heat and deprived of their ability to catalyze reactions are said to be *denatured*.

pH

Enzymes also function best at a particular pH. For most enzymes, the optimal pH is at or near a pH of 7:

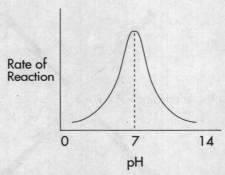

Other enzymes operate at a low pH. For instance, pepsin, the digestive enzyme found in the stomach, is most effective at an extremely acidic pH of 2.

Here's another important piece of information:

Most enzymes are active over a narrow range of pH.

ENZYME REGULATION

We know that enzymes control the rates of chemical reactions. But what regulates the activity of enzymes? It turns out that a cell can control enzymatic activity by regulating the conditions that influence the shape of the enzyme. For example, some enzymes have **allosteric sites**, a region of the enzyme other than the active site to which a substance can bind. Substances called **allosteric regulators** can either inhibit or activate enzymes. An **allosteric inhibitor** will bind to an allosteric site and keep the enzyme in its inactive form while an **allosteric activator** will bind to an enzyme and induce its active form. Allosteric enzymes are subject to feedback inhibition in which the formation of an end product inhibits an earlier reaction in the sequence.

Most enzymes can be inhibited by certain chemical substances. If the substance has a shape that fits the active site of an enzyme, it can compete with the substrate and effectively inactivate the enzyme. This is called **competitive inhibition**. Usually a competitive inhibitor is structurally similar to the normal substrate. In **noncompetitive inhibition**, the inhibitor binds with the enzyme at a site other than the active site and inactivates the enzyme by altering its shape. This prevents the enzyme from binding with the substrate at the active site.

ADENOSINE TRIPHOSPHATE (ATP)

We've all heard the expression that nothing is for free. The same holds true in nature. Here's a fundamental principle about energy that we must address:

> Energy cannot be created or destroyed. In other words, the sum of energy in the universe is constant.

This rule is called the **first law of thermodynamics**. As a result, the cell cannot take energy out of thin air. Rather, it must harvest it somewhere.

The **second law of thermodynamics** states that energy transfer leads to less organization. That means the universe tends toward disorder (or **entropy**).

As we just saw, everything an organism does requires energy. How then can the cell acquire the energy it needs without becoming a major mess? Fortunately, it's through **adenosine triphosphate (ATP)**.

ATP, as the name indicates, consists of a molecule of adenosine bonded to three phosphates. The great thing about ATP is that an enormous amount of energy is packed into those phosphate bonds, particularly the third bond:

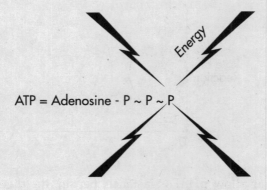

$$\text{ATP} = \text{Adenosine} - P \sim P \sim P$$

When a cell needs energy, it takes one of these potential-packed molecules of ATP and splits off the third phosphate, forming adenosine diphosphate (ADP) and one loose phosphate (P_i), while releasing energy in the process:

$$\text{ATP} \rightarrow \text{ADP} + P_i + \text{energy}$$

The energy released from this reaction can then be put to whatever use the cell so pleases. None of this, of course, means that the cell is above the laws of thermodynamics. But within those constraints, ATP is the best source of energy the cell has available. It is relatively neat (only one bond needs to be broken to release that energy) and relatively easy to form. As such, it's the ideal energy currency for living things. Anywhere you look in nature, you're bound to find ATP.

SOURCES OF ATP

But where does all this ATP come from? It is produced in one of two ways: (1) through photosynthesis, or (2) through cellular respiration.

Photosynthesis involves the transformation of solar energy into chemical energy. The overall reaction is:

$$6CO_2 + 6H_2O + \text{sunlight} \rightarrow C_6H_{12}O_6 + 6O_2$$

Plants use their leaves to capture sunlight and make glucose. As we saw in Chapter 2, organisms that make their own food are known as **autotrophs**, or **producers**, whereas those that gain their energy through the ingestion of others are known as **heterotrophs**.

We'll hold off on our discussion of photosynthesis until Chapter 5. For now, let's focus on the other means of ATP production—**cellular respiration**.

THE SHORTHAND VERSION

In **cellular respiration**, ATP is produced through the breakdown of nutrients. You'll recall from the beginning of this chapter that many organic molecules are important to cells because they are energy-rich. This is where that energy comes into play.

In the shorthand version, cellular respiration looks something like this:

$$C_6H_{12}O_6 + 6O_2 \rightarrow 6CO_2 + 6H_2O + ATP$$

Notice that we've taken a sugar, perhaps a molecule of glucose, and combined it with oxygen and water to produce carbon dioxide, water, and energy in the form of our old friend, ATP. However, as you probably already know, the actual picture of what really happens is far more complicated.

Generally speaking, we can break cellular respiration down to two different approaches: aerobic respiration or anaerobic respiration. If ATP is made in the presence of oxygen, we call it **aerobic respiration**. If oxygen isn't present, we call it anaerobic respiration. Let's jump right in with aerobic respiration.

AEROBIC RESPIRATION

Aerobic respiration consists of four stages: (1) **glycolysis**, (2) **formation of acetyl CoA**, (3) the **Krebs cycle**, and (4) **oxidative phosphorylation**. There are so many steps within each stage that some students find this topic too confusing to follow. Don't sweat it. We've come up with a simple method to keep all the stages of cellular respiration in order: Just keep track of the number of carbons at each stage.

Stage 1: Glycolysis

The first stage begins with **glycolysis**, the splitting (-*lysis*) of glucose (*gluco*-). Glucose is a six-carbon molecule that is broken into two three-carbon molecules, called **pyruvic acid**. This breakdown of glucose also results in the net production of two molecules of ATP:

$$\text{Glucose} + 2\text{ ATP} + 2\text{NAD}^+ \rightarrow 2\text{ Pyruvic acid} + 4\text{ ATP} + 2\text{NADH}$$

Although we've written glycolysis as if it were a single reaction, this process doesn't occur in one step. In fact, it requires a sequence of reactions!

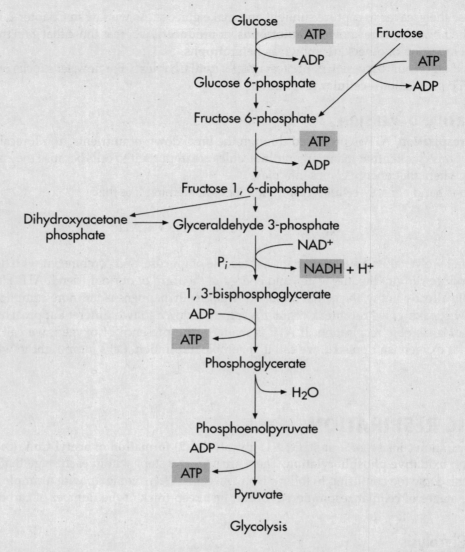

Glycolysis

Fortunately, you don't need to memorize these steps for the test. What you do need to know is that glucose doesn't *automatically* generate ATP. It has to be activated. Once glucose is "powered up," it eventually splits into two pyruvic acids.

If you take a good look at the reaction above, you'll see two ATPs are needed to produce four ATPs. You've probably heard the expression, "You have to spend money to make money." In biology, you have to invest ATP to make ATP: Our investment of two ATPs yielded four ATPs, for a net gain of two.

A second product in glycolysis is 2 NADH, which results from the transfer of H^+ to the hydrogen carrier NAD^+. NADH will be used elsewhere in respiration to make additional ATP.

There are two important tidbits to remember regarding glycolysis:

- Occurs in the cytoplasm
- Net of 2 ATPs produced
- 2 pyruvic acids formed
- 2 NADH produced

Once the cell has undergone glycolysis, it has two options: It can continue anaerobically, or it can switch to true aerobic respiration. As we'll soon see, the cell's decision has a lot to do with the environment in which it finds itself. If oxygen is present, many cells switch directly to aerobic respiration. If no oxygen is present, those same cells may carry out anaerobic respiration. Still others have no choice, and carry out only anaerobic respiration, with or without oxygen.

Since ETS is more likely to ask you about aerobic respiration, we'll look closely at the remaining steps.

However, before we do so, let's jump back to those important organelles, the **mitochondria**. We already know from our discussion in Chapter 3 that the mitochondria are the sites of cellular respiration. Now it's time to see exactly *where* they manufacture ATP.

The double membrane of the mitochondria divides the organelle into four regions:

- The **matrix**

- The **inner mitochondrial membrane**

- The **intermembrane space**

- The **outer membrane**

Let's have a look:

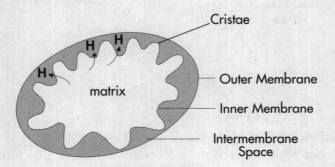

Why do you need to know about the different regions within a mitochondrion? Because we'll soon see that several of the stages of aerobic respiration occur within these regions of the mitochondria—and ETS loves to ask about you questions about where things occur! Keep them in mind. We'll be discussing them below.

Stage 2: Formation of Acetyl CoA
When oxygen is present, pyruvic acid enters the mitochondrion. Each pyruvic acid (a three-carbon molecule) is converted to **acetyl coenzyme A** (a two-carbon molecule) and CO_2 is released:

$$2 \text{ Pyruvic acid} + 2 \text{ Coenzyme A} + 2NAD^+ \rightarrow 2 \text{ Acetyl CoA} + 2 CO_2 + 2NADH$$

Are you keeping track of our carbons? We've now gone from a three-carbon molecule to a two-carbon molecule. The extra carbons leave the cell in the form of CO_2. Once again, two molecules of NADH are also produced.

Stage 3: The Krebs Cycle
The next stage is the **Krebs cycle**, also known as the **citric acid cycle**. Each of the two acetyl coenzyme A molecules will enter the Krebs cycle, one at a time, and all the carbons will ultimately be converted to CO_2. This stage occurs in the matrix of the mitochondria.

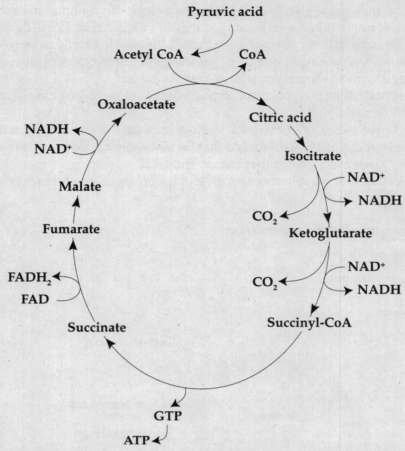

Pyruvic acid

Acetyl CoA ← CoA

Oxaloacetate

NADH ←
NAD⁺

Malate

Fumarate

FADH₂ ←
FAD

Succinate

Citric acid

Isocitrate

NAD⁺
→ NADH

CO_2 ←

Ketoglutarate

NAD⁺
→ NADH

CO_2 ←

Succinyl-CoA

GTP

ATP ←

Let's track the carbons again. Each molecule of acetyl CoA produced from the second stage of aerobic respiration combines with **oxaloacetate**, a four-carbon molecule, to form a six-carbon molecule, **citric acid** or "citrate":

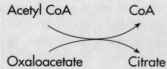

Acetyl CoA   CoA

Oxaloacetate   Citrate

Since the cycle begins with a four-carbon molecule, oxaloacetate, it also has to end with a four-carbon molecule to maintain the cycle. So how many carbons do we have to lose to keep the cycle going? Two carbons, both of which will be released as CO_2. Now the cycle is ready for another turn with the second acetyl CoA.

With each turn of the cycle, three additional types of molecules are produced:

- 1 ATP
- 3 NADH
- 1 FADH₂

To figure out the total number of products per molecule of glucose, we simply double the number of products—after all, we started off the Krebs cycle with two molecules of acetyl CoA for each molecule of glucose!

Now we're ready to tally up the number of ATP produced.

After the Krebs cycle, we've made only four ATP—two ATP from glycolysis and two ATP from the Krebs cycle.

Although that seems like a lot of work for only four ATP, we have also produced hydrogen carriers in the form of NADH and $FADH_2$. These molecules will in turn produce lots of ATP.

Stage 4: Oxidative Phosphorylation

We said earlier that ATP is the energy currency of the cell. While this is true, ATP is not the only molecule that stores energy. Sometimes energy is stored by electron carriers like NAD^+ and FAD.

As electrons (and their accompanying hydrogens) are removed from a molecule of glucose, they carry with them much of the energy that was originally stored in their chemical bonds. These electrons—and their accompanying energy—are then transferred to readied hydrogen carrier molecules. In the case of cellular respiration, these charged carriers are NADH and $FADH_2$.

Let's see how many electron carriers we've produced. We now have:

- Two NADH molecules from glycolysis

- Two NADH from the production of acetyl CoA

- Six NADH from the Krebs cycle

- Two $FADH_2$ from the Krebs cycle

That gives us 12 "loaded" electron carriers altogether.

Now let's consider the fate of all the electrons removed from the breakdown of glucose. Here's what happens. The electron carriers—NADH and $FADH_2$—now "shuttle" electrons to the electron transport chain.

First, the hydrogen atoms are split into hydrogen ions and electrons:

$$H_2 \rightarrow 2\ H^+ + 2\ e^-$$

Two interesting things now happen. First, the high-energy electrons from NADH and $FADH_2$ are passed down a series of protein carrier molecules that are embedded in the cristae, the membrane along the inner membrane. Some of these are iron-containing carriers called **cytochromes**. Take a look at the entire chain:

Electron flow

Each carrier molecule hands down the electrons to the next molecule in the chain. The electrons travel down the electron transport chain until they reach the final electron acceptor, oxygen. Oxygen combines with these electrons (and some hydrogens) to form water. This explains the "aerobic" in aerobic respiration. If oxygen weren't available to accept the electrons, the last carrier in the chain would be "stuck" with them, thereby shutting down the whole process of ATP production.

CHEMIOSMOSIS

At the same time that electrons are being passed down the electron transport chain, another mechanism is at work. Remember those hydrogen ions (also called *protons*) that split off from the original hydrogen atom? Some of the energy released from the electron transport chain is used to pump hydrogen ions across the inner mitochondrial membrane to the intermembrane space. The pumping of hydrogen ions into the intermembrane space creates a **pH gradient**, or **proton gradient**. The potential energy established in this gradient is responsible for the production of ATP.

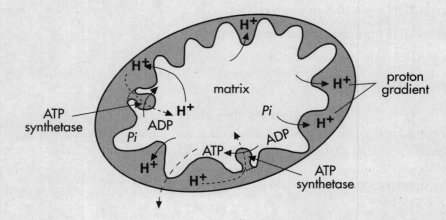

These hydrogen ions can diffuse across the inner membrane only by passing through channels called **ATP synthase**. Meanwhile, ADP and P_i are on the other side of these protein channels. The flow of protons through these channels produces ATP by combining ADP and P_i on the matrix side of the channel. This process is called **oxidative phosphorylation**.

Three other things you're expected to know for the AP Biology Exam:

- Every NADH yields 3 ATP (except NADH from glycolysis produces 2 ATP)
- Every FADH$_2$ yields 2 ATP
- The total number of ATP produced in Stage 4 is 32 ATP

Don't concern yourself with the exact number of NADH, FADH$_2$, CO$_2$, and ATP produced at each stage. For the AP Biology Exam, all you need to remember are the big steps and the overall outcome:

STAGES OF AEROBIC RESPIRATION		
Step in Respiration	**Takes Place in the ...**	**Net Result**
Glycosis	Cytoplasm	2ATP (+2NADH)
Split to acetyl CoA	Cytoplasm	0ATP (+2NADH)
Krebs cycle	Mitochondrial	2ATP (6NADH + 2FADH$_2$)
Oxidative Phosphorylation	Inner mitochondrial membrane	32ATP
		Net: 36 ATP

ANAEROBIC RESPIRATION

Some organisms can't undergo aerobic respiration. They're anaerobic. They can't use oxygen to make ATP. We just learned that oxygen is important because it's the final electron acceptor in the electron transport chain.

How do anaerobic organisms derive energy? Since glycolysis is an anaerobic process, they can make 2 ATP from this stage. However, instead of carrying out the other stages of aerobic respiration (the Krebs cycle, electron transport chain, and oxidative phosphorylation), these organisms carry out a process called **fermentation**. Under anaerobic conditions, pyruvic acid is converted to either **lactic acid** or **ethyl alcohol** (or **ethanol**) and carbon dioxide.

For the AP Biology Exam, you should remember the two pathways anaerobic organisms undergo: glycolysis and fermentation. Unfortunately, anaerobic respiration is not very efficient. It only results in a gain of 2 ATP for each molecule of glucose broken down.

As you can see from the chart below, there are two basic end products in anaerobic respiration. In both pathways, the NADH formed during glycolysis reduces (adds hydrogen to) pyruvate.

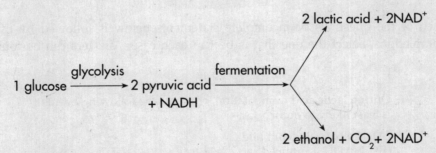

What types of organisms undergo fermentation?

- Yeast cells and some bacteria make ethanol and carbon dioxide.

- Other bacteria produce lactic acid.

YOUR MUSCLE CELLS CAN FERMENT

Although human beings are aerobic organisms, they can actually carry out fermentation in their muscle cells. Have you ever had a cramp? If so, that cramp was the consequence of anaerobic respiration.

When you exercise, your muscles require a lot of energy. To get this energy, they convert enormous amounts of glucose to ATP. But as you continue to exercise, your body doesn't get enough oxygen to keep up with the demand in your muscles. This creates an oxygen debt. What do your muscle cells do? They switch over to anaerobic respiration. Pyruvic acid produced from glycolysis is converted to lactic acid. As a consequence, the lactic acid causes your muscles to ache.

KEY WORDS

enzymes
organic catalysts
exergonic reactions
endergonic reactions
activation energy
enzyme specificity
substrates
active site
enzyme-substrate complex
induced fit
coenzymes
cofactors
allosteric sites
allosteric regulators
allosteric inhibitors

allosteric activator
competitive inhibition
noncompetitive inhibition
first law of thermodynamics
second law of thermodynamics
entropy
adenosine triphosphate (ATP)
photosynthesis
autotrophs (or producers)
heterotrophs
cellular respiration
aerobic respiration
anaerobic respiration
glycolysis

pyruvic acid
mitochondria
acetyl coenzyme A
Krebs cycle (or citric acid cycle)
oxaloacetate
citric acid
cytochromes
hydrogen ion gradient
    (or proton gradient)
ATP synthase
oxidative phosphorylation
fermentation
lactic acid
ethanol (or ethyl alcohol)

QUIZ

<u>Directions:</u> Each of the questions or incomplete statements below is followed by five suggested answers or completions. Select the one that is best in each case. Answers can be found on pages 237–238.

1. During lactic acid fermentation, which of the following events is least likely to occur?

    (A) An oxygen debt builds
    (B) NADH is reduced
    (C) Lactate accumulates in the muscle tissue
    (D) Production of ATP
    (E) NAD⁺ is recycled and returns to the Krebs cycle

2. All of the following statements about exergonic reactions are true EXCEPT:

   (A) The products contain less energy than the reactants.
   (B) The energy released can be used to perform work.
   (C) They usually occur instantaneously.
   (D) They release energy in the form of ATP or heat.
   (E) They require energy input before the reaction can proceed.

3. Which of the following statements is true concerning enzymes?

   (A) They always require a coenzyme.
   (B) They become hydrolyzed during a chemical reaction.
   (C) They are consumed in the reaction.
   (D) They operate under a narrow pH.
   (E) They are polymers of carbohydrates.

4. The final acceptor of electrons in the electron transport chain is:

   (A) hydrogen
   (B) water
   (C) oxygen
   (D) NADH
   (E) $FADH_2$

5. Enzymes are affected by all of the following EXCEPT:

   (A) pH
   (B) temperature
   (C) chemical agents
   (D) concentration of substrates
   (E) concentration of water

6. The reaction below represents what type of reaction?

   $C_6H_{12}O_6 + 6O_2 \rightarrow 6CO_2 + 6H_2O + energy$

   (A) Aerobic respiration
   (B) Anaerobic respiration
   (C) Glycolysis
   (D) Fermentation
   (E) Photosynthesis

7. All of the following stages are considered aerobic processes EXCEPT:

   (A) the Krebs cycle
   (B) formation of acetyl CoA
   (C) glycolysis
   (D) electron transport chain
   (E) oxidative phosphorylation

8. Glucose is labeled with ^{14}C and followed as it is broken down to produce CO_2, H_2O, and ATP in a human muscle cell. This label will be detectable in which of the following sites?

(A) In the cytoplasm only
(B) In the mitochondria only
(C) In the cytoplasm and the mitochondria
(D) In the nucleus and the mitochondria
(E) In the nucleus, cytoplasm, and the mitochondria

9. Eukaryotes carry out oxidative phosphorylation in the mitochondrion. The analogous structure used by prokaryotes to carry out oxidative phosphorylation is the:

(A) nuclear membrane
(B) cell wall
(C) plasma membrane
(D) ribosome
(E) endoplasmic reticulum

Directions: Each group of questions consists of five lettered headings followed by a list of numbered phrases or sentences. For each numbered phrase or sentence, select the one heading that is most closely related to it and fill in the corresponding oval on the answer sheet. Each heading may be used once, more than once, or not at all in each group.

Questions 10–14

(A) Pyruvic acid
(B) Alcoholic fermentation
(C) FAD^+
(D) ATP synthase
(E) Cofactor

10. An oxidized coenzyme

11. A method of anaerobic respiration

12. An end product of glycolysis

13. An enzyme that utilizes the proton-motive force of chemiosmosis

14. A compound that plays a role in the catalytic action of an enzyme

5

Photosynthesis

Plants and algae are producers. All they have to do is bask in the sun, churning out the glucose necessary for life.

As we saw in Chapter 4, photosynthesis is the process by which light energy is converted to chemical energy. Here's an overview of **photosynthesis**:

$$6CO_2 + 6H_2O \xrightarrow{\text{sunlight}} C_6H_{12}O_6 + 6O_2$$

You'll notice from this equation that carbon dioxide and water are the raw materials plants use in manufacturing simple sugars. But remember, there's much more to photosynthesis than what's shown in the simple reaction above. We'll soon see that this beautifully orchestrated process occurs thanks to a whole host of special enzymes and pigments. But before we turn to the stages in photosynthesis, let's talk about where photosynthesis occurs.

THE ANATOMY OF A LEAF

Photosynthesis occurs in the leaves of plants. Here's a cross-sectional view of a typical leaf:

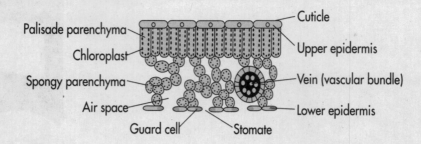

If you look closely at the leaf of any plant, the first thing you'll notice is the waxy covering called the cuticle. The **cuticle**, which is visible even without the cross-section above, is produced by the upper epidermis to protect the leaf from water loss through evaporation. Just below the **upper epidermis** is the **palisade parenchyma**. These cells contain lots of **chloroplasts**, which are the primary sites of photosynthesis.

Now let's look at an individual chloroplast. If you split the membrane of a chloroplast, you'll find a fluid-filled region called the **stroma**. Inside the stroma are structures that look like stacks of coins. These structures are the **grana**.

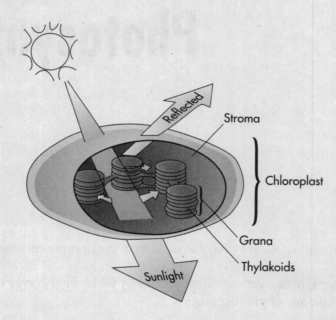

The many disk-like structures that make up the grana are called the **thylakoids**, and they contain chlorophyll, the light-absorbing pigment and enzymes that drives photosynthesis.

Now we'll discuss some structures that are not involved in photosynthesis but are important for the AP test. Just below the palisade parenchyma, you'll find irregular-shaped cells called **spongy parenchyma**. The **conducting tissues** are found in this layer of the leaf. The vascular bundles include **xylem** and **phloem**, "vessels" that transport materials throughout the plant. At the **lower epidermis** are tiny openings called **stomates,** which allow for gas exchange and transpiration. Surrounding each stomate are **guard cells**, which control the opening and closing of the stomates.

A CLOSER LOOK AT PHOTOSYNTHESIS

There are two stages in photosynthesis: the **light reaction** (also called the light dependent reaction) and the **dark reaction** (also called the light independent reaction). The whole process begins when the **photons** (or "energy units") of sunlight strike a leaf, activating chlorophyll and exciting electrons. The activated chlorophyll molecule then passes these excited electrons down to a series of electron carriers, ultimately producing ATP and NADPH. The whole point of the light reaction is to produce two things: (1) energy in the form of ATP and (2) electron carriers, specifically NADPH.

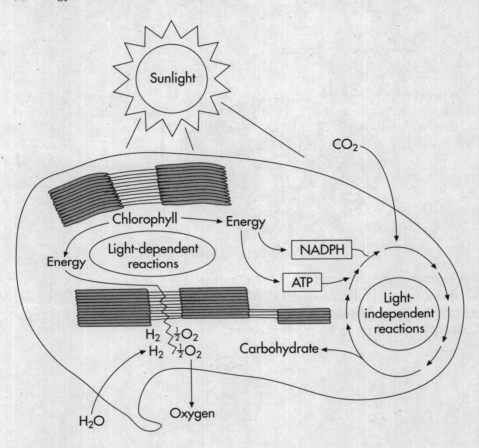

Both of these products, along with carbon dioxide, are then used in the dark reaction (light independent) to make carbohydrates.

THE LIGHT REACTION

Many light-absorbing pigments participate in photosynthesis. Some of the more important ones are **chlorophyll *a***, **chlorophyll *b***, and **carotenoids**. These pigments are clustered in the thylakoid membrane into units called antenna complexes.

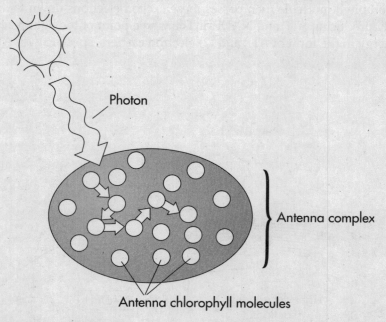

Photon

Antenna complex

Antenna chlorophyll molecules

All of the pigments within a unit are able to "gather" light, but they're not able to "excite" the electrons. Only one special molecule—located in the **reaction center**—is capable of transforming light energy to chemical energy. In other words, the other pigments, called **antenna pigments**, "gather" light and "bounce" the energy to the reaction center.

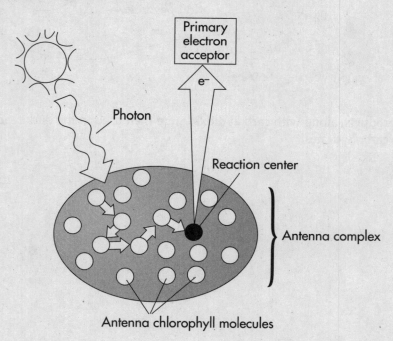

Primary electron acceptor

e⁻

Photon

Reaction center

Antenna complex

Antenna chlorophyll molecules

There are two types of reaction centers: **photosystem I (PS I)** and **photosystem II (PS II)**. The principal difference between the two is that each reaction center has a type of chlorophyll—**chlorophyll *a***—that absorbs a particular wavelength of light. For example, **P680**, the reaction center of photosystem II, has a maximum absorption at a wavelength of 680 nanometers. The reaction center for photosystem I—**P700**—best absorbs light at a wavelength of 700 nanometers.

When this process forms **ATP**, it is called **photophosphorylation**. We're using light (that's *photo*) and ADP and phosphates (that's *phosphorylation*) to produce ATP.

Noncyclic photophosphorylation

The noncyclic method produces ATP using both photosystem I and photosystem II. When a leaf captures sunlight, the energy is sent to P680, the reaction center for photosystem II. The activated electrons are trapped by P680 and passed to a molecule called the primary acceptor and then passed down to carriers in the electron transport chain and eventually enter photosystem I. Some of the energy that dissipates as electrons move along the chain of acceptors will be used to "pump" protons across the membrane into the thylakoid lumen. When the reaction center P680 absorbs light, it also splits water into oxygen, hydrogen ions, and electrons. That process is called **photolysis**. The electrons from photolysis replace the missing electrons in photosystem I.

Remember the chemiosmotic theory mentioned in aerobic respiration? Well, the same mechanism applies to photosynthesis. Here's how it works: Hydrogen ions accumulate inside the thylakoids when photolysis occurs. A proton gradient is established. As the hydrogen ions move through ATP synthase, ATP and P_i produce ATP.

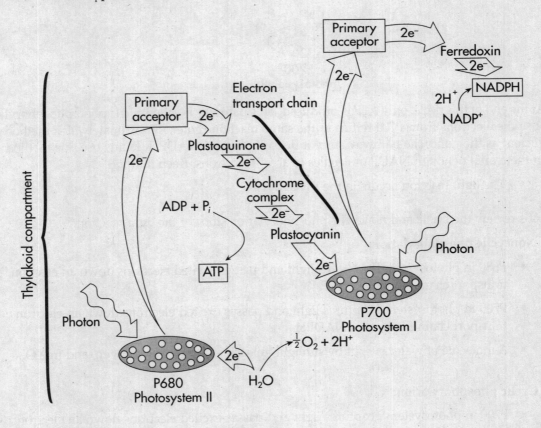

When these electrons in photosystem I receive a second boost, they're activated again. The electrons are passed through a second electron transport chain until they reach the final electron acceptor **NADP$^+$ to make NADPH**. Since the system can't just "give away" electrons, it needs to find a way to replace the electrons it loses.

Cyclic photophosphorylation

The cyclic method uses a much simpler pathway to generate ATP. The electrons in photosystem I are excited and leave the reaction center, P700. They are passed from carrier to carrier in the electron transport system and eventually return to P700:

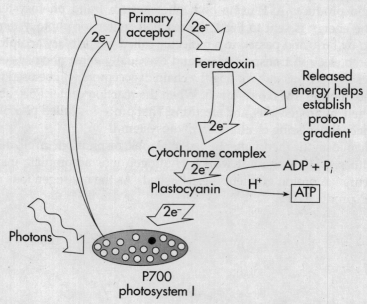

At the end of this cycle, only ATP is produced. This pathway is called cyclic photophosphorylation because the electrons from P700 return to the same reaction center. Unfortunately, this method isn't as efficient as the noncyclic pathway since it doesn't produce NADPH. Plants use this method only when there aren't enough NADP molecules to accept electrons. Keep in mind:

- The light reaction occurs in the thylakoids.

Let's review the cyclic and noncyclic phosphorylation steps of the light reaction:

Noncyclic phosphorylation:

- P680 in photosystem II captures light and passes excited electrons down an electron transport chain to produce ATP.

- P700 in photosystem I captures light and passes excited electrons down an electron transport chain to produce NADPH.

- A molecule of water is split by sunlight, releasing electrons, hydrogen, and free O_2.

Cyclic phosphorylation:

- P700 in photosystem I captures light and passes excited electrons down an electron transport chain to produce ATP.

- NADPH is *not* produced.
- Water is *not* split by sunlight.

Both reactions occur in the grana of the leaf, where the thylakoids are found. One more thing: The light-absorbing pigments and enzymes for the light-dependent reactions are found within the thylakoids.

THE LIGHT INDEPENDENT REACTION

Now let's turn to the dark reaction. The dark reaction uses the products of the light reaction—ATP and NADPH—to make sugar. We now have energy to make glucose, but what do plants use as their carbon source? CO_2, of course. You've probably heard of the term **carbon fixation**. All this means is that CO_2 from the air is converted into carbohydrates. This step occurs in the stroma of the leaf.

The Calvin cycle: The C_3 pathway

We're finally ready to make glucose. CO_2 enters the **Calvin cycle** and combines with a 5–carbon molecule called **ribulose bisphosphate (RuBP)** to make an unstable six-carbon compound. The enzyme RuBP carboxylase or *rubisco* catalyzes this reaction.

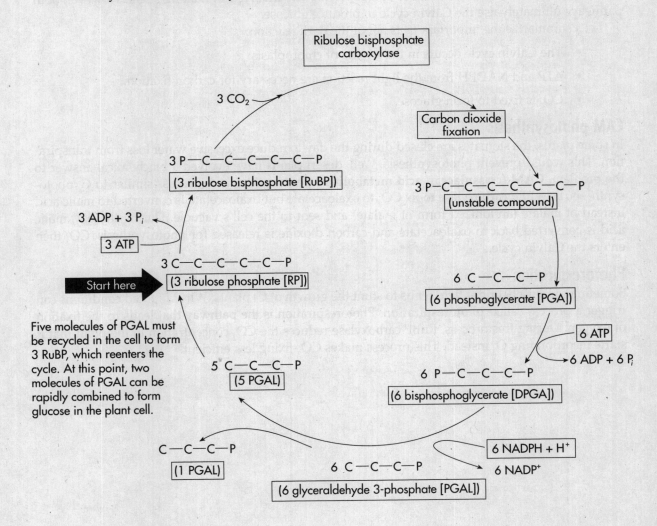

The easiest way to view the Calvin cycle is to consider 6RuBP and 6CO_2 at the start. Next, 12 ATP and 12 NADPH are used to convert 12 PGA to 12 PGAL, an energy-rich molecule. ADP and $NADP^+$ are released and then recycled into the thylakoid where they will again be available for the light-dependent reactions. Two of the PGAL are used to make glucose while the remaining 10 are rearranged into 6 RuBPs ready for the next round of the cycle. Since G3P (PGAL), a three-carbon molecule, is the first stable product, this method of producing glucose is called the C_3 **pathway**.

C_4: The alternative pathway

The C_3 pathway is not the only way to "fix" CO_2. Some plants, such as sugar cane and corn, have a more efficient way to fix carbon dioxide. In these plants, carbon dioxide first "combines" with **phosphoenolpyruvate (PEP)** in mesophyll cells to form oxaloacetate, a four-carbon molecule. The enzyme that fixes PEP, **PEP carboxylase**, has a high affinity for CO_2 even under unusually low concentrations. Oxaloacetate is then converted to malate. Malate enters the bundle sheath cells, a tissue surrounding the leaf vein, and is converted to pyruvate and carbon dioxide. Carbon dioxide is then released for uptake into the regular Calvin cycle to make glucose.

The **C_4 pathway** works particularly well for plants found in hot, dry climates. It enables them to fix CO_2 even when the supply is greatly diminished. For our purposes, just remember that the subscripts in both C_3 and C_4 refer to the number of carbons *initially* involved in making sugar. However, both pathways ultimately use the Calvin cycle to produce glucose.

Let's summarize the important facts about the dark reaction:

- The Calvin cycle occurs in the stroma of chloroplasts.

- ATP and NADPH from the light reaction are necessary for carbon fixation.

- CO_2 is fixed to form glucose.

CAM photosynthesis

In some plants, the stomates are closed during the day to reduce excessive water loss from transpiration. This would prevent photosynthesis. Well, desert plants have evolved a biochemical answer to the problem. **CAM (crassulacean acid metabolism) photosynthesis**, a process similar to C_4 photosynthesis, uses PEP carboxylase to fix CO_2 to oxaloacetate but oxaloacetate is converted to malic acid instead of malate (an ionized form of malate) and sent to the cell's vacuole. During the day, malic acid is converted back to oxaloacetate and carbon dioxide is released for photosynthesis. CO_2 then enters the Calvin cycle.

Photorespiration

Sometimes intensely bright light tends to stunt the growth of C_3 plants. Why? Lighted conditions can trigger a process called **photorespiration**. Photorespiration is the pathway that leads to the fixation of oxygen. During this process, RuBP carboxylase reduces the CO_2 concentration to the point that it starts incorporating O_2 instead. This process makes CO_2-fixing less efficient.

KEY WORDS

photosynthesis
cuticle
upper epidermis
palisade parenchyma
chloroplasts
stroma
grana
thylakoids
spongy parenchyma
conducting tissues
lower epidermis
stomates
guard cells
light reaction (light dependent reaction)
dark reaction (light independent reaction)
photons
chlorophyll a
chlorophyll b
carotenoids
reaction center

antenna pigments
photosystem I (PS I)
photosystem II (PS II)
P680
P700
photophosphorylation
ATP
noncyclic method of photophosphorylation
cyclic method of photophosphorylation
NADPH
photolysis
carbon fixation (or C3 pathway)
Calvin cycle
ribulose bisphosphate (RuBP)
phosphoenolpyruvate (PEP)
C_4 pathway
photorespiration
CAM photosynthesis

QUIZ

<u>Directions:</u> Each of the questions or incomplete statements below is followed by five suggested answers or completions. Select the one that is best in each case. Answers can be found on page 238.

1. All of the following statements are correct regarding the light reaction in photosynthesis EXCEPT:

   (A) The antennae pigments capture sunlight.
   (B) The reaction center in photosystem II is P680.
   (C) Light energy is converted to chemical energy.
   (D) The electrons are activated and passed along an electron transport chain.
   (E) Carbon dioxide is used to make glucose.

2. When electrons are passed along the electron transport chain during noncylic photophosphorylation, in photosystem I, this results in the production of

   (A) NADPH
   (B) ATP
   (C) cytochromes
   (D) water
   (E) glucose

3. One of the outcomes of noncyclic photophosphorylation that does not occur in cyclic photophosphorylation is

  (A) light energy is absorbed
  (B) ATP is produced
  (C) NADPH is produced
  (D) electrons are passed along the electron transport chain
  (E) electrons trapped by photosystem I are energized by sunlight

4. Which of the following processes requires sunlight, ADP, and water?

  (A) Photosystem I
  (B) Photosystem II
  (C) Glycolysis
  (D) Calvin cycle
  (E) Electron transport system

5. The primary difference between the C_3 and C_4 pathway is that

  (A) C_4 has special pigments involved in photosynthesis.
  (B) C_3 plants use CO_2 to make glucose whereas C_4 plants use O_2.
  (C) C_3 plants fix carbon during the day whereas C_4 plants fix carbon during the night.
  (D) C_4 plants have a higher rate of photosynthesis than C_3 plants.
  (E) C_3 plants are better adapted to intense sunlight than C_4 plants.

Directions: Each group of questions consists of five lettered headings followed by a list of numbered phrases or sentences. For each numbered phrase or sentence, select the one heading that is most closely related to it and fill in the corresponding oval on the answer sheet. Each heading may be used once, more than once, or not at all in each group.

Questions 6–9

  (A) Stroma
  (B) Thylakoids
  (C) Photolysis
  (D) Carotenoids
  (E) Ribulose bisphosphate

6. Contains enzymes of the carbon-fixing reactions

7. Contains enzymes of photophosphorylation

8. The five-carbon molecule involved in the dark reaction

9. The molecule that accepts CO_2 during the Calvin cycle

6

Molecular Genetics

DNA: THE BLUEPRINT OF LIFE

All living things possess an astonishing degree of organization. For the simplest single-celled organism as for the largest mammal, millions of reactions and events must be precisely coordinated for life to exist. This coordination is directed from the nucleus of the cell, by **deoxyribonucleic acid**, or **DNA**.

> DNA is the hereditary blueprint of the cell.

The DNA of a cell is contained in structures called chromosomes. The chromosomes are DNA wrapped around proteins called histones. When the genetic material is in a loose form throughout the nucleus it is called **euchromatin**. When the genetic material is fully condensed into coils it is called **heterochromatin**. Situated in the nucleus, chromosomes direct and control all the processes necessary for life, including passing themselves, and their information, on to future generations. In this chapter, we look at precisely how they accomplish this.

THE MOLECULAR STRUCTURE OF DNA

The DNA molecule consists of two strands that wrap around each other to form a long, twisted ladder called a **double helix**. The structure of DNA was brilliantly deduced in 1956 by two scientists named Watson and Crick.

STRUCTURE OF DNA

DNA is made up of repeated subunits of **nucleotides**. Each nucleotide has a **five-carbon sugar, a phosphate**, and **a nitrogenous base**. Take a look at the nucleotide below. This particular nucleotide contains a nitrogenous base called **adenine**:

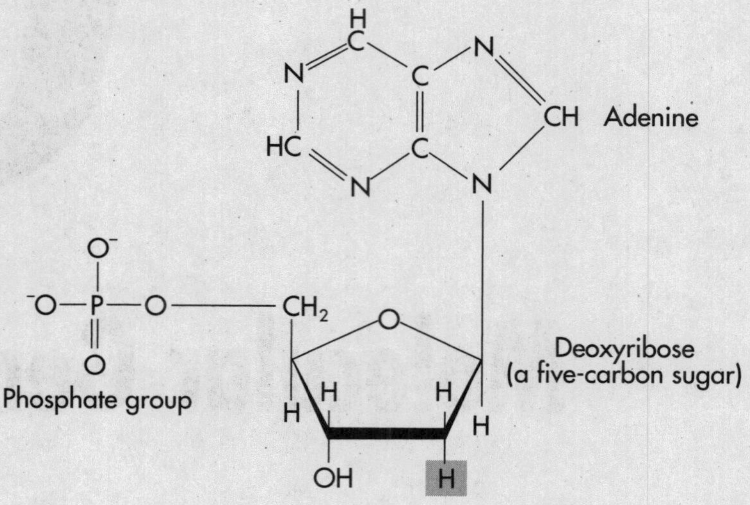

The name of the pentagon-shaped sugar in DNA is **deoxyribose**. Hence, the name *deoxyribo*nucleic acid. Notice that the sugar is linked to two things: a phosphate and a nitrogenous base. A nucleotide can have one of four different nitrogenous bases:

- **Adenine**—a purine (double-ringed nitrogenous base)
- **Guanine**—a purine (double-ringed nitrogenous base)
- **Cytosine**—a pyrimidine (single-ringed nitrogenous base)
- **Thymine**—a pyrimidine (single-ringed nitrogenous base)

Any of these four bases can attach to the sugar. As we'll soon see, this is extremely important when it comes to the "sense" of the genetic code in DNA.

The nucleotides can link up in a long chain to form a single strand of DNA. Here's a small section of a DNA strand:

The nucleotides themselves are linked together by **phosphodiester bonds**.

Two DNA Strands

Now let's look at the way two DNA strands get together. Again, think of DNA as a ladder. The sides of the ladder consist of alternating sugar and phosphate groups, while the rungs of the ladder consist of a pair of nitrogenous bases:

Hydrogen bonds

The nitrogenous bases pair up in a particular way. Adenine in one strand always binds to thymine (A-T or T-A) in the other strand. Similarly, guanine always binds to cytosine (G-C or C-G). This predictable matching of the bases is known as **base pairing**.

The two strands are said to be **complementary**. This means that if you know the sequence of bases in one strand, you'll know the sequence of bases in the other strand. For example, if the base sequence in one DNA strand is A-T-C, the base sequence in the complementary strand will be T-A-G.

The two DNA strands run in opposite directions. You'll notice from the figure above that each DNA strand has a 5' end and a 3' end, so-called for the carbon that ends the strand (i.e., the fifth carbon in the sugar ring is at the 5' end, while the third carbon is at the 3' end). The 5' end has a phosphate group and the 3' end has an OH, or "hydroxyl," group. The 5' end of one strand is always opposite to the 3' end of the other strand. The strands are therefore said to be **antiparallel**.

The DNA strands are linked by **hydrogen bonds**. Two hydrogen bonds hold adenine and thymine together and three hydrogen bonds hold cytosine and guanine together.

Before we go any further, let's review the base pairing in DNA:

- Adenine pairs up with thymine (A-T or T-A) by forming *two* hydrogen bonds.

- Cytosine pairs up with guanine (G-C or C-G) by forming *three* hydrogen bonds.

WHY DNA IS IMPORTANT

You'll recall from our discussion of bioenergetics in Chapter 4 that enzymes are proteins that are essential for life. This is true not only because they help liberate energy stored in chemical bonds, but also because they direct the construction of the cell. This is where DNA comes into the picture.

DNA's main role is directing the manufacture of proteins. These proteins, in turn, regulate everything that occurs in the cell. But DNA does not *directly* manufacture proteins. Instead, DNA passes its information to an intermediate molecule known as **ribonucleic acid (RNA)**. These RNA molecules carry out the instructions in DNA, producing the proteins that determine the course of life.

The flow of genetic information is therefore:

$$DNA \rightarrow RNA \rightarrow proteins$$

This is the central doctrine of molecular biology.

We said in the beginning of this chapter that DNA is the hereditary blueprint of the cell. By directing the manufacture of proteins, DNA serves as the cell's blueprint. But how is DNA inherited? For the information in DNA to be passed on, it must first be copied. This copying of DNA is known as **DNA replication**.

DNA REPLICATION

Since the DNA molecule is twisted over on itself, the first step in replication is to unwind the double helix by breaking the hydrogen bonds. This is accomplished by an enzyme called **helicase**. The exposed DNA strands now form a *y*-shaped **replication fork**:

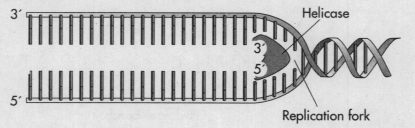

Now each strand can serve as a template for the synthesis of another strand. DNA replication begins at specific sites called **origins of replication**. Since the DNA helix twists and rotates during DNA replication, another class of enzymes, called DNA **topoisomerases**, cuts and rejoins the helix to prevent tangling. The enzyme that performs the actual addition of nucleotides alongside the naked strand is **DNA polymerase**. But DNA polymerase, oddly enough, can only add nucleotides to the end of an existing strand. Therefore, before it can add nucleotides, DNA polymerase needs **RNA primase**, the enzyme that initiates the whole process using an RNA primer. This molecule adds the first few nucleotides to each new DNA strand, starting off replication.

One strand is called the **leading strand**, and it is made continuously. That is, the nucleotides are steadily added one after the other by DNA polymerase. The other strand—the **lagging strand**—is made discontinuously. Unlike the leading strand, the lagging strand is made in pieces of nucleotides known as **Okazaki fragments**. Why is the lagging strand made in small pieces?

You'll recall from the diagrams above that a molecule of DNA has two ends: a 3' end and a 5' end. Normally, nucleotides are added only in the 5' to 3' direction. However, when the double-helix is "unzipped," one of the two strands is oriented in the opposite direction—3' to 5'. Since DNA polymerase doesn't work in this direction, nucleotides—the Okazaki fragments—need to be added in pieces. These fragments are eventually linked together by the enzyme **DNA ligase** to produce a continuous strand. Finally, hydrogen bonds form between the new base pairs, leaving two identical copies of the original DNA molecule.

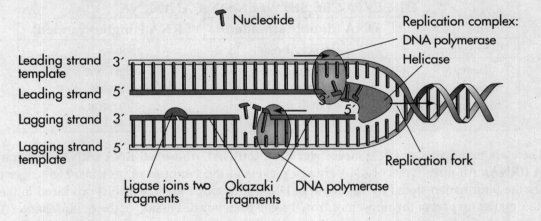

When DNA is replicated, we don't end up with two entirely new molecules. Each new molecule has half of the original molecule. Since DNA replicates in this way, by conserving half of the original molecule in each of the two new ones, it is said to be **semiconservative**.

Many enzymes and proteins are involved in DNA replication. The ones you'll need to know for the AP Biology Exam are DNA helicase, DNA polymerase, DNA ligase, topoisomerase, and RNA primer:

- **Helicase** (unwinds our double helix into two strands).
- **Polymerase** (adds nucleotides to an existing strand).
- **Ligase** (brings together the Okazaki fragments).
- **Topoisomerase** (cuts and rejoins the helix).
- **RNA primer** (RNA nucleotides used to provide 3' end to which DNA polymerase can add nucleotides.

RNA

Now that we've seen how *DNA* is replicated, let's take a look at how the genetic code is expressed as proteins. Genetic information is first passed to intermediates called RNA. Here's a "roadmap" of how information is transferred from DNA to proteins:

$$DNA \xrightarrow[\text{in the nucleus}]{\text{transcription}} RNA \xrightarrow[\text{in the cytoplasm}]{\text{Translation}} Proteins$$

Before we discuss what RNA does, let's talk about its structure. Although RNA is also a nucleotide, it differs from DNA in three ways:

1. RNA is single-stranded, not double-stranded.

2. The five-carbon sugar in RNA is **ribose** instead of deoxyribose.

3. The RNA nitrogenous bases are adenine, guanine, cytosine, and a new base called **uracil**. Uracil replaces thymine.

Here's a table to compare DNA and RNA. Keep these differences in mind—ETS loves to test you on them:

DIFFERENCES BETWEEN DNA AND RNA		
	DNA (double stranded)	**RNA (single stranded)**
Sugar:	deoxyribose	ribose
Bases:	adenine guanine cytosine thymine *	adenine guanine cytosine uracil *

There are three types of RNA: **messenger RNA (mRNA)**, **ribosomal RNA (rRNA)**, and **transfer RNA (tRNA)**. All three types of RNA are key players in the synthesis of proteins. Messenger RNA copies the information stored in the strand of DNA. Ribosomal RNA, which is produced in the nucleolus, makes up part of the ribosomes. You'll recall from our discussion of the cell in Chapter 3 that the ribosomes are the sites of protein synthesis. We'll see how they function a little later on. The third type of RNA, transfer RNA, shuttles amino acids to the ribosomes. Transfer RNA is responsible for bringing the appropriate amino acids into place at the appropriate time. It does this by reading the message carried by the mRNA. Now that we know about the different types of RNA, let's see how they direct the synthesis of proteins.

PROTEIN SYNTHESIS

Protein synthesis involves three basic steps: **transcription, RNA processing**, and **translation.**

TRANSCRIPTION

Transcription involves copying the genetic code directly from DNA. The initial steps in transcription are similiar to the initial steps in DNA replication. The obvious difference is that whereas in replication we end up with a complete copy of the cell's DNA, in translation we end up with only a partial copy in the form of mRNA.

Transcription involves three phases: initiation, elongation, and termination. As in DNA replication, the first initiation step in transcription is to unwind and unzip the DNA strands using helicase. Transcription begins at special sequences of the DNA strand called **promoters**. Since RNA is single-stranded, we have to copy only *one* of the two DNA strands. The strand that serves as the template is the **sense strand**. The other strand that lies dormant is the **antisense strand**:

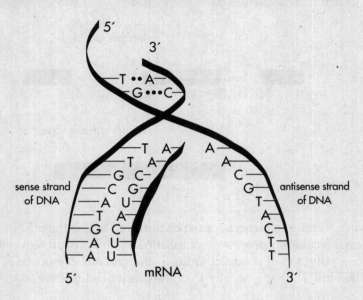

This time, RNA nucleotides line up alongside one DNA strand to form an mRNA strand. A new enzyme, **RNA polymerase**, brings free-floating RNA nucleotides to the DNA strand. As we saw earlier, guanines and cytosines pair up. However, the RNA molecule replaces the DNA's thymine molecules with uracils. In other words, the exposed adenines are now paired up with uracil instead of thymine.

Once the mRNA finishes adding on nucleotides and reaches a termination point, it separates from the DNA template, completing the process of transcription. The new RNA has now transcribed, or "copied," the sequence of nucleotide bases directly from the exposed DNA strand.

RNA Processing

Now the mRNA strand is ready to move out of the nucleus. But before the mRNA molecule can leave the nucleus, it must be **processed**. That means the mRNA has to be modified before it exits the nucleus.

A newly made mRNA molecule (called the heterogenous nuclear RNA) contains more nucleotides than it needs to code for a protein. The mRNA consists of both coding regions and noncoding regions. The regions that express the code for the polypeptide are **exons**. The noncoding regions in the mRNA are **introns**.

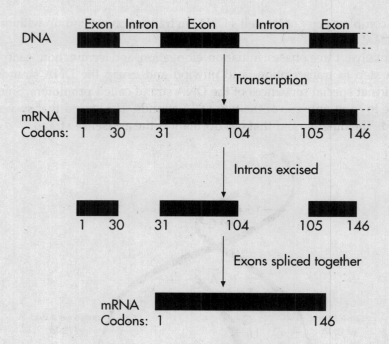

The introns—the intervening sequences—must be removed before the mRNA leaves the nucleus. The removal of introns is a complex process accomplished by an RNA-protein complex called a **spliceosome**. In addition, a **poly(A) tail** is added to the 3' end and a **5' cap** is added to the 5' end. This process produces a final mRNA that is shorter than the transcribed mRNA.

TRANSLATION

Now, the mRNA leaves the nucleus and searches for a ribosome. The mRNA molecule carries the message from DNA in the form of **codons**, a group of three bases, or "letters," that corresponds to one of 20 amino acids. The genetic code is redundant, meaning that certain amino acids are specified by more than one codon.

The mRNA attaches to the ribosome and "waits" for the appropriate amino acids to come to the ribosome. That's where tRNA comes in. A tRNA molecule has a unique three-dimensional structure, which resembles a four-leaf clover:

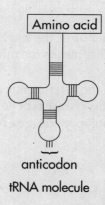

One end of the tRNA carries an amino acid. The other end, called an **anticodon**, has three nitrogenous bases that can base pair with the codon in the mRNA.

Transfer RNAs are the "go-betweens" in protein synthesis. Each tRNA becomes charged and enzymatically attaches to an amino acid in the cell's cytoplasm and "shuttles" it to the ribosome. The charging enzymes involved in forming the bond between the amino acid and tRNA require ATP.

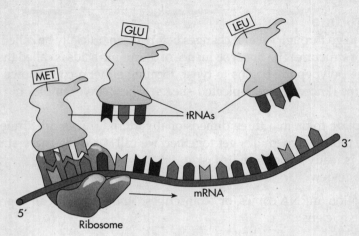

Translation also involves three phases: initiation, elongation, and termination. Initiation begins when a ribosome attaches to the mRNA.

What does the ribosome do? It helps the process along by holding everything in place while the RNAs assist in assembling polypeptides.

Initiation

Ribosomes contain two binding sites: an **A site** and a **P site**. An **initiator tRNA**, often met tRNA, serves as the initial activation of translation and occupies the P site. Once the initiator tRNA is in place, a second tRNA can attach to the A site.

For example, the tRNA with the complementary anticodon, U-A-C, is methionine's personal shuttle: It carries no other amino acids. Every time the mRNA has an A-U-G, it is certain to pick up a methionine.

Elongation

The addition of amino acids is called elongation. Remember that the mRNA contains many codons or "triplets" of nucleotide bases. As each amino acid is brought to the mRNA, it is linked to its neighboring amino acid by a **peptide bond**. When many amino acids link up, a **polypeptide** is formed.

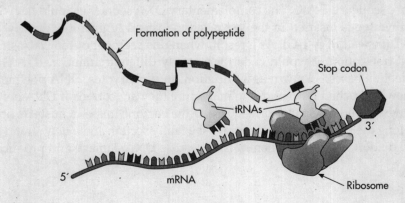

Termination

How does this process know when to stop? The synthesis of a polypeptide is ended by stop codons. A codon doesn't always code for an amino acid; there are three that serve as a stop codon. For example, U-A-A, U-A-G, and U-G-A are all **stop codons**. Termination occurs when the ribosome runs into one of these three stop codons.

Higher Protein Structure

The polypeptide has to go through several changes before it can officially be called a protein. **Proteins** can have four levels of structure. The linear sequence of the amino acids is called the **primary structure** of a protein. Now the polypeptide begins to twist, forming either a coil (known as an alpha helix), or zig-zagging patterns (known as beta-pleated sheets). These are examples of proteins' **secondary structures**.

Next, the polypeptide folds in a three dimensional pattern. This is called the **tertiary structure**. Finally, when two or more polypeptides get together, we call it a **quaternary structure**. Once that's complete the protein is ready to perform its task.

How about a little review?

- In transcription, mRNA copies, or "transcribes," the code from an exposed strand of DNA in the nucleus.

- The mRNA is "processed" by having its introns, or noncoding sequences, removed.

- Now, ready to be translated, mRNA proceeds to the ribosome.

- Free-floating amino acids are picked up by tRNA and shuttled over to the ribosome, where mRNA awaits.

- In translation, the anticodon of a tRNA molecule carrying the appropriate amino acid base pairs with the codon on the mRNA.

- As new tRNA molecules match up to new codons, the ribosomes hold them in place, allowing peptide bonds to form between the amino acids.

- The newly formed polypeptide grows until a stop codon is reached.

- The polypeptide or protein is released into the cell.

MUTATIONS

What happens if errors occur during DNA replication? Suppose a short segment of a template DNA strand has the base sequence T-A-C. The complementary DNA segment should be A-T-G. If, for some reason, another thymine is inserted in the last position instead of guanine, the new strand of DNA would read A-T-T (instead of A-T-G). As a result, when this segment of the DNA strand underwent transcription and translation, it would code for a totally different amino acid. A change in one or more of the nucleotide bases is known as a **mutation**. In prokaryotes, DNA polymerase proofreads (or checks) the newly attached nucleotides to make sure they are correct. If DNA polymerase finds a mistake, it will go back and insert the correct base. If the enzyme misses a mistake, other DNA repair enzymes move along the DNA strand and remove the faulty section and a new strand is inserted. Mutagens, such as radiation and certain chemicals, can cause DNA damage and produce a mutation.

Base Substitution

One of the simplest types of mutation is a base substitution mutation. It results in a change in only one pair of nucleotides. A base substitution can destroy the function of a protein if a codon is altered and produces a different amino acid (**missense mutation**) or if a base substitution produces a termination codon (**nonsense mutation**). Other times, a base substitution produces a closely related amino acid. This type of mutation may be undetectable (**silent mutation**).

GENE REGULATION

What controls gene transcription and how does an organism express only certain genes? Most of what we know about gene regulation comes from our studies of *E. coli.* In bacteria, the region of bacterial DNA that regulates gene expression is called an **operon**. One of the best-understood operons is the *lac* operon, which controls expression of the enzymes that break down lactose.

The operon consists of four major parts: structural genes, the regulatory gene, the promoter gene, and the operator.

- **Structural genes** are genes that code for enzymes needed in a chemical reaction. These genes will be transcribed at the same time to produce particular enzymes. In the *lac* operon, three enzymes (beta galactosidase, galactose permease, and thiogalactoside transacetylase) involved in digesting lactose are coded for.

- The **promoter gene** is the region where the RNA polymerase binds to begin transcription.

- The **operator** is a region that controls whether transcription will occur.

- The **regulatory gene** codes for a specific regulatory protein called the *repressor*. The repressor is capable of attaching to the operator and blocking transcription. If the repressor binds to the operator, transcription will not occur. On the other hand, if the repressor does not bind to the operator, RNA polymerase moves right along the operator and transcription occurs. In the *lac* operon, the **inducer**, lactose, binds to the repressor, causing it to fall off the operator, and "turns on" transcription.

Other operons, such as the *trp* operon, operate in a similar manner except that mechanism is continually "turned on" and is only "turned off" in the presence of high levels of the amino acid, tryptophan. Tryptophan is a product of the pathway that codes for the *trp* operon. When tryptophan combines with the *trp* repressor protein, it causes the repressor to bind to the operator, which turns the operon "off" thereby blocking transcription. In other words, a high level of tryptophan acts to repress the further synthesis of tryptophan.

RECOMBINANT DNA

Recombinant DNA is a hybrid of DNA from two or more sources. For example, DNA with a gene of interest from eukaryotic organisms can be transferred into a bacterial genome. The branch of technology that produces new organisms or products by transferring genes between cells is called **genetic engineering**.

How does this procedure work? To form a recombinant DNA, first, a plasmid, a small circular DNA, is cut into pieces by a restriction enzyme. **Restriction enzymes** are enzymes that cut DNA at a specific site. Then the same restriction enzyme is used on the DNA of interest. All fragments produced by the enzyme will now have the same **sticky ends** and the fragments from the gene of interest can be inserted into the plasmid. The new product, now called a **recombinant plasmid** or a vector DNA, can self-replicate in its host.

The DNA fragments can be separated according to their molecular weight using **gel electrophoresis**. Electrophoresis of DNA fragments separates them by size. The smaller the fragments the faster they move through the gel. Restriction enzymes are also used to create a molecular fingerprint. These **DNA fingerprints**, which give the precise patterns of DNA fragments, are accurate and already play a major role in identifying criminals.

KEY WORDS

deoxyribonucleic acid (DNA)
double helix
nucleotides
nitrogenous base
deoxyribose
adenine
guanine
cytosine
thymine
phosphodiester bonds
base pairing
complementary
antiparallel
hydrogen bonds
ribonucleic acid (RNA)
DNA replication
helicase
replication fork
topoisomerase
origins of replication
DNA polymerase
RNA primase
leading strand
lagging strand
Okazaki fragments
DNA ligase
semiconservative
ribose
uracil
mRNA
rRNA
tRNA
purine
pyrimidine
protein synthesis

transcription
translation
promoters
sense strand
antisense strand
RNA polymerase
processed
exons
introns
poly(A) tail
5' cap
codons
anticodon
peptide bond
stop codons
proteins
primary structure
secondary structure
tertiary structure
quaternary structure
mutation
base substitution
missense mutation
nonsense mutation
silent mutation
operon
structural genes
promoter gene
operator
regulatory gene
inducer
genetic engineering
restriction enzyme
recombinant plasmid
gel electrophoresis
DNA fingerprinting

QUIZ

Directions: Each of the questions or incomplete statements below is followed by five suggested answers or completions. Select the one that is best in each case. Answers can be found on page 239.

1. Which of the following cellular organelles is most closely associated with the transcription activity of RNA?

   (A) Mitochondria
   (B) Nucleus
   (C) Ribosomes
   (D) Golgi apparatus
   (E) Lysosome

2. The correct sequence between genes and their phenotypic expression is

   (A) RNA-DNA-protein-trait
   (B) DNA-RNA-protein-trait
   (C) Protein-DNA-RNA-trait
   (D) trait-DNA-RNA-protein
   (E) trait-protein-DNA-RNA

3. Which of the following substances is found in RNA molecules but not in DNA molecules?

   (A) Adenine
   (B) Phosphate group
   (C) Thymine
   (D) Deoxyribose
   (E) Ribose

4. If a messenger RNA codon is UAC, which of the following would be the complementary anticodon triplet in the transfer RNA?

   (A) ATG
   (B) AUC
   (C) AUG
   (D) ATT
   (E) ATC

5. After posttranscriptional modification, the polypeptide from a eukaryotic cell typically undergoes substantial alteration that results in

   (A) excision of introns
   (B) addition of a poly(A) tail
   (C) formation of peptide bonds
   (D) a change in the overall conformation of a polypeptide
   (E) the synthesis of amino acids

6. All of the following enzymes are involved in the replicative process of DNA EXCEPT:

(A) DNA helicase
(B) DNA polymerase
(C) RNA polymerase
(D) RNA primase
(E) DNA ligase

7. Which of the following represents the maximum number of amino acids that could be incorporated into a polypeptide encoded by 21 nucleotides of messenger RNA?

(A) 3
(B) 7
(C) 21
(D) 42
(E) 63

Directions: Each group of questions consists of five lettered headings followed by a list of numbered phrases or sentences. For each numbered phrase or sentence, select the one heading that is most closely related to it and fill in the corresponding oval on the answer sheet. Each heading may be used once, more than once, or not at all in each group.

Questions 8–11

(A) Okazaki fragments
(B) Transposons
(C) Lagging strand
(D) Leading strand
(E) hnRNA

8. DNA segments that can move around the genome

9. The strand that is first assembled in discrete nucleotide segments

10. The discontinuous strand during DNA replication

11. Unprocessed RNA molecules that are precursors

Cell Reproduction

CELL DIVISION

Every second, thousands of cells are dying throughout our bodies. Fortunately, the body replaces them at an amazing rate. In fact, **epidermal**, or skin, cells die off and are replaced so quickly that the average 18-year-old grows an entirely new skin every few weeks. The body keeps up this unbelievable rate thanks to the mechanisms of **cell division**.

This chapter takes a closer look at how cells divide. But remember, cell division is only a small part of the life cycle of a cell. Most of the time, cells are busy carrying out their regular activities. Since we covered DNA replication in chapter 6, let's now look at how cells pass this genetic material to their offspring.

THE CELL CYCLE

Every cell has a life cycle—the period from the beginning of one division to the beginning of the next. The cell's life cycle is known as the **cell cycle**. The cell cycle is divided into two periods: **interphase** and **mitosis**. Let's look at the cell cycle of a typical cell:

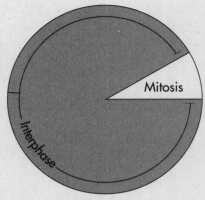

Notice that most of the life of a cell is spent in interphase.

INTERPHASE: THE GROWING PHASE

Interphase is the time span from one cell division to another. We call this stage interphase (*inter-* means between) because the cell has not yet started to divide. Although biologists sometimes refer to interphase as the "resting stage," the cell is definitely not inactive. This phase is when the cell carries out its regular activities. All the proteins and enzymes it needs to grow are produced during interphase.

The three stages of interphase

Interphase can be divided into three stages: **G1**, **G2**, and **S phases**:

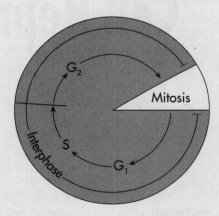

The most important phase is the **S phase**. That's when the cell replicates its genetic material. The first thing a cell has to do before undergoing mitosis is to duplicate all of its chromosomes. During interphase, every single chromosome in the nucleus replicates.

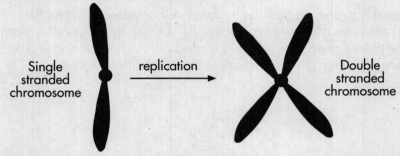

Single stranded chromosome — replication → Double stranded chromosome

You'll notice that the original chromosome and its duplicate are still linked, like Siamese twins. These identical chromosomes are now called **sister chromatids** (each individual structure is called a chromatid). Take a look:

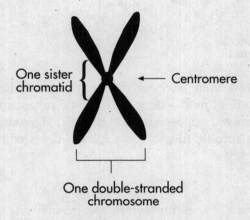

One sister chromatid — Centromere

One double-stranded chromosome

The chromatids are held together by a structure called the **centromere**. Although the chromosomes have been duplicated, they are still considered a single unit. Once duplication has been done, we're ready for the big breakup: mitosis.

We've already said that replication occurs during the S phase of interphase, so what happens during G1 and G2? During these stages, the cell produces proteins and enzymes. For example, during G1 the cell produces all of the enzymes required for DNA replication (as we saw in Chapter 6, that means DNA helicase, DNA polymerase, and DNA ligase). By the way, "G" stands for "gap," but we can also associate it with "growth."

Let's recap:

- The cell cycle consists of two things: interphase and mitosis.

- During the S stage of interphase, the chromosomes replicate.

- Growth and preparation for mitosis occurs during the G1 and G2 stages of interphase.

MITOSIS: THE DANCE OF THE CHROMOSOMES

Once the chromosomes have replicated, the cell is ready to begin mitosis. Mitosis is the period when the cell divides. Mitosis consists of a sequence of four stages: prophase, metaphase, anaphase, and telophase.

STAGE 1: PROPHASE

One of the first signs of prophase is the disappearance of the nucleolus. In prophase, the chromosomes thicken, forming coils upon coils, and become visible. (During interphase, the chromosomes are not visible. Rather, the genetic material is scattered throughout the nucleus and is called **chromatin**. It is only during prophase that we can properly speak about the chromosomes.)

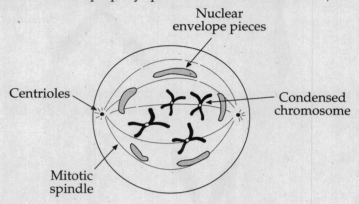

Now the cell has plenty of room to "sort out" the chromosomes. Remember centrioles? During prophase, these cylindrical bodies found within microtubule organizing centers (MTOCs) start to move away from each other, toward opposite ends of the cell. The centrioles will spin out a system of microtubules known as the **spindle fibers**. These spindle fibers will attach to a structure in the centromere called **kinetochores**.

STAGE 2: METAPHASE

The next stage is called metaphase. The chromosomes now begin to line up along the equitorial plane of the cell.

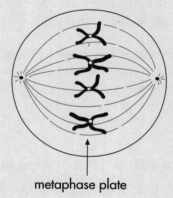

metaphase plate

That's because the spindle fibers that are attached to the kinectochores of the chromosomes.

STAGE 3: ANAPHASE

During anaphase, the sister chromatids of each chromosome separate at the centromere and migrate to opposite poles because microtubules attached to the kinetochores pull each one apart. The chromosomes are pulled apart at the centromere by the microtubules, which begin to shorten.

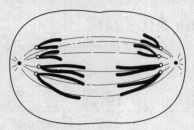

Each half of a pair of sister chromatids now moves to opposite poles of the cell.

STAGE 4: TELOPHASE

The final phase of mitosis is telophase. A nuclear membrane forms around each set of chromosomes and the nucleoli reappear.

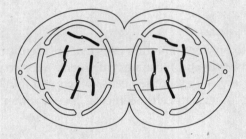

The nuclear membrane is ready to divide. Now it's time to split the cytoplasm in a process known as **cytokinesis**. Look at the figure below and you'll notice that the cell has begun to split along a **cleavage furrow** (which is produced by actin microfilaments):

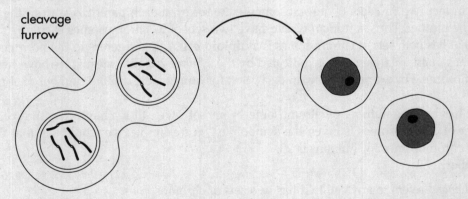

cleavage furrow

A cell membrane forms about each cell and they split into two distinct daughter cells. The division of the cytoplasm yields two daughter cells.

Here's one thing to remember: Cytokinesis occurs differently in plant cells. The cell doesn't form a cleavage furrow. Instead, a partition called a **cell plate** forms down the middle region.

Stage 5: Interphase

Once the daughter cells are produced, they reenter the initial phase—interphase—and the whole process starts over. The cell goes back to its original state. Once again, the chromosomes become invisible, and the genetic material is called chromatin.

But How Will I Remember All That?

For mitosis, you may already have your own **mnemonic**. If not, here's a table with a mnemonic we created for you.

IPMAT	
Interphase	**I** is for **Interlude**
Prophase	**P** is for **Prepare**
Metaphase	**M** is for **Meet**
Anaphase	**A** is for **Apart**
Telophase	**T** is for **Tear**

Purpose of Mitosis

Mitosis has two purposes: (1) to produce daughter cells that are identical copies of the parent cell, and (2) to maintain the proper number of chromosomes from generation to generation. For our purposes, mitosis occurs in just about every cell except sex cells. When you think of mitosis, remember: "Like begets like." Hair cells "beget" other hair cells; skin cells "beget" other skin cells, etc.

HAPLOIDS VERSUS DIPLOIDS

Every organism has a certain number of chromosomes. For example, fruit flies have 8 chromosomes, humans have 46 chromosomes, and dogs have 78 chromosomes. It turns out that most eukaryotic cells have in fact two full sets of chromosomes—one set from each parent: one from the father and one from the mother. Thus, a human would have two sets of 23 chromosomes each.

A cell that has both sets of chromosomes is a **diploid cell**. Diploid refers to the normal number of chromosomes in an organism and is indicated by the symbol 2n. That means, we have two copies of each chromosome. For example, we would say that for humans the diploid number of chromosomes is 46.

If a cell has half the number of chromosomes—one set—we call it a **haploid cell**. A cell with half the number of chromosomes is given the symbol "n." For example, we would say that the haploid number of chromosomes for humans is 23.

Remember:

- *Diploid* refers to any cell that has two sets of chromosomes.

- *Haploid* refers to any cell that has one set of chromosomes.

Why do we need to know the terms *haploid* and *diploid*? Because they are extremely important when it comes to sexual reproduction. As we've seen, 46 is the normal diploid number for human beings. We can say, therefore, that human cells have 46 chromosomes. However, this isn't entirely correct.

Human chromosomes come in pairs called **homologues**. So while there are 46 of them altogether, there are actually only 23 *distinct* chromosomes. The **homologous chromosomes**, which make up each pair, are similar in size and shape and express similar traits. This occurs in all sexually reproducing organisms. In fact, this is the essence of sexual reproduction: Each parent donates half its chromosomes to its offspring.

GAMETES

Although most cells in the human body are diploid (i.e., filled with pairs of chromosomes), there are special cells that are haploid (i.e., unpaired). These haploid cells are called **sex cells**, or **gametes**. Why do we have haploid cells?

If each gamete has the same number of chromosomes as the parental cell, then the zygote produced would have 46 pairs instead of 23.

But, by splitting the diploid cell into a haploid cell known as a gamete, each parent contributes one-half his or her full complement. This production of gametes is **meiosis**.

AN OVERVIEW OF MEIOSIS

To preserve the diploid number of chromosomes in an organism, each parent must contribute only half of its chromosomes. This is the point of meiosis. Since sexually reproducing organisms need only haploid cells for reproduction, meiosis is limited to sex cells in special sex organs called **gonads**. In males, the gonads are the **testes**, while in females they are the **ovaries**. The special cells in these organs—also known as **germ cells**—produce haploid cells (n), which then combine to restore the diploid (2n) number during fertilization:

$$\text{female gamete (n) + male gamete (n) = zygote (2n)}$$

When it comes to genetic variation, meiosis is a big plus. Variation, in fact, is the driving force of evolution. The more variation there is in a population, the more likely it is that some members of the population will survive extreme changes in the environment. Meiosis is far more likely to produce these sorts of variations than is mitosis, and therefore confers selective advantage on sexually reproducing organisms. We'll come back to this theme in Chapter 12.

A CLOSER LOOK AT MEIOSIS

Meiosis actually involves two rounds of cell division called **meiosis I** and **meiosis II**.

Before meiosis begins, the diploid cell goes through interphase. Just as in mitosis, double-stranded chromosomes are formed during this phase.

MEIOSIS I

Meiosis I consists of four stages: **prophase I**, **metaphase I**, **anaphase I**, and **telophase I**.

Prophase I

Prophase I is a little more complicated than regular prophase. As in mitosis, the nuclear membrane disappears, the chromosomes become visible, and the centrioles move to opposite poles of the nucleus. But that's where the similarity ends.

The major difference involves the movement of the chromosomes. In meiosis, the chromosomes line up side-by-side with their counterparts (homologues). This event is known as **synapsis**.

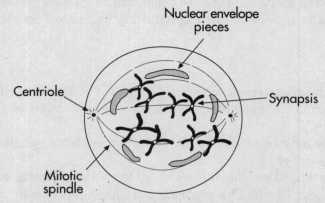

Synapsis involves two sets of chromosomes that come together to form a **tetrad** (or a **bivalent**). A tetrad consists of four chromatids. Synapsis is followed by **crossing-over**, the exchange of segments between homologous chromosomes.

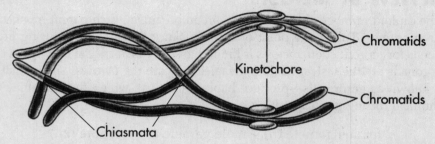

What's unique in prophase I is that "pieces" of chromosomes are exchanged between the homologous partners. This is one of the ways organisms produce genetic variation. By the end of prophase I, the chromosomes will have exchanged regions containing several **alleles**, or different forms of the same gene. By the end of prophase, the homologous chromosomes are held together only at specialized regions called **chiasmata**.

Metaphase I

As in mitosis, the chromosome pairs—now called tetrads—line up at the metaphase plate. By contrast, you'll recall that in regular metaphase the chromosomes lined up individually.

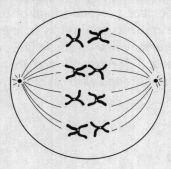

Anaphase I

During anaphase I, one of each pair of chromosomes within a tetrad separates and moves to opposite poles. Notice that the chromosomes do not separate at the centromere. They separate with their centromeres intact.

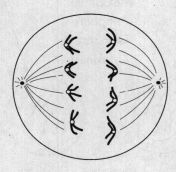

The chromosomes now go on to their respective poles.

Telophase I

During telophase I, the nuclear membrane forms around each set of chromosomes.

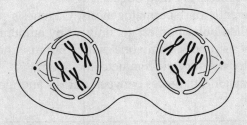

Finally, the cells undergo cytokinesis, leaving us with two daughter cells. Notice that at this point the nucleus contains the haploid number of chromosomes, but each chromosome is a duplicated chromosome.

MEIOSIS II

The purpose of the second meiotic division is to separate the duplicated chromosomes, and is virtually identical to mitosis. Let's run through the steps in meiosis II.

After a brief period, the cell undergoes a second round of cell division. During prophase II, the chromosomes once again condense and become visible. In metaphase II, the chromosomes move toward the metaphase plate. This time they line up single file, not as pairs. During anaphase II , the chromatids of each chromosome split at the centromere and are pulled to opposite ends of the cell. At telophase II, a nuclear membrane forms around each set of chromosomes and a total of *four* haploid cells are produced:

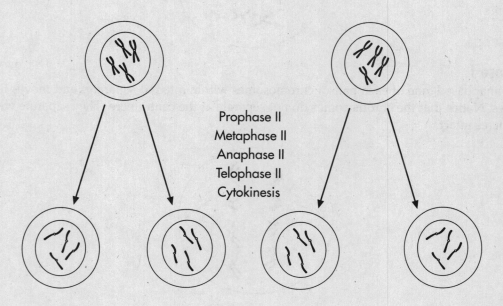

Prophase II
Metaphase II
Anaphase II
Telophase II
Cytokinesis

GAMETOGENESIS

Meiosis is also known as gametogenesis. If sperm cells are produced then meiosis is called **spermatogenesis**. During spermatogenesis, four sperm cells are produced for each diploid cell. If an egg cell or an ovum is produced, this process is called oogenesis.

Oogenesis is a little different from spermatogenesis. Oogenesis produces only one ovum, not four. The other three cells, called **polar bodies**, get only a tiny amount of cytoplasm and eventually degenerate. Why does oogenesis produce only one ovum? Because the female wants to conserve as much cytoplasm as possible for the surviving gamete, the ovum.

Here's a summary of the major differences between mitosis and meiosis:

MITOSIS	MEIOSIS
• Occurs in somatic (body) cells	• Occurs in germ (sex) cells
• Produces identical cells	• Produces gametes
• Diploid cell → diploid cells	• Diploid cell → haploid cells

MUTATIONS

Mutations also occur at the chromosome level. Sometimes, a set of chromosomes has an extra or a missing chromosome. This occurs because of **nondisjunction**—the chromosomes failed to separate properly during meiosis. This error, which produces the wrong number of chromosomes in a cell, results in severe genetic defects. For example, humans have 23 pairs of chromosomes. An individual with **Down's syndrome** has three—instead of two—copies of the 21st chromosome.

Chromosomal abnormalities also occur if one or more segments of a chromosome break. The most common examples are **translocation** (a segment of a chromosome moves to another chromosome), **inversion** (a segment of a chromosome is inserted in the reverse orientation), **deletion** (a segment of a chromosome is lost), and **duplication** (an extra copy of a chromosomal segment is introduced).

Here's an example of a translocation:

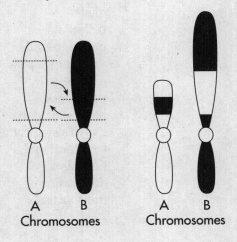

Examples of translocation are **transposons**, DNA segments that have the ability to move around the genome. Sometimes when they move, they leave behind mutations and they can cause mutations by inserting into a gene.

Fortunately, in most cases, damaged DNA can usually be repaired with special repair enzymes.

KEY WORDS

cell division

cell cycle

mitosis

interphase

S phase

G1 phase

G2 phase

sister chromatids

centromere

prophase

chromatin

metaphase

anaphase

telophase

spindle fibers

kinetochores

metaphase plate

cytokinesis

cleavage furrow

cell plate

diploid cell

haploid cell

homologues

homologous chromosomes

sex cell or gametes

meiosis

gonads

testes

ovaries

germ cells

synapsis

tetrad

crossing-over

alleles

chiasmata

gametogenesis

spermatogenesis

oogenesis

polar bodies

nondisjunction

Down's syndrome

translocation

inversion

duplication

transposons

QUIZ

<u>Directions:</u> Each of the questions or incomplete statements below is followed by five suggested answers or completions. Select the one that is best in each case. Answers can be found on page 240.

1. If the diploid number for an organism is 24, the haploid number is

   (A) 6
   (B) 12
   (C) 18
   (D) 24
   (E) 48

2. When homologous chromosomes come into close contact during meiosis I, this is called

   (A) crossing-over
   (B) synapsis
   (C) tetrad
   (D) cytokinesis
   (E) interphase

3. During meiosis there are two rounds of all of the following stages EXCEPT

   (A) prophase
   (B) metaphase
   (C) anaphase
   (D) telophase
   (E) interphase

4. All of the following represent events that occur with a cell during meiosis I EXCEPT

   (A) Homologous chromosomes become closely associated during synapsis
   (B) Sister chromatids disjoin and chromosomes move to opposite poles
   (C) Tetrads are held together at chiasmata
   (D) The nucleolus disappears and chromatin condenses
   (E) Genetic material is exchanged between nonsister chromatids

Directions: Each group of questions consists of five lettered headings followed by a list of numbered phrases or sentences. For each numbered phrase or sentence, select the one heading that is most closely related to it and fill in the corresponding oval on the answer sheet. Each heading may be used once, more than once, or not at all in each group.

Questions 5–9

(A) G2 phase
(B) Chromatin
(C) Centromere
(D) Centrioles
(E) S phase

5. DNA-protein complex making up eukaryotic chromosomes

6. Period during which DNA is replicated

7. Structure that holds chromatids together

8. Contained within microtubule organizing centers

9. Period of renewed protein synthesis

8

Heredity

GREGOR MENDEL: THE FATHER OF GENETICS

What is genetics? In its simplest form, genetics is the study of heredity. It explains how certain characteristics are passed on from parents to children. Much of what we know about genetics was discovered by the monk **Gregor Mendel** in the 19th century. Since then, the field of genetics has vastly expanded. As scientists study the mechanisms of genetics, they've developed new ways of manipulating genes. For example, scientists have isolated the gene that makes insulin, a human hormone, and now use bacteria to make large quantities of it. But before we get ahead of ourselves, let's study the basic rules of genetics.

Let's begin then with some of the fundamental points of genetics:

- Each **trait**—or expressed characteristic—is produced by hereditary factors known as **genes**. Within a chromosome, there are many genes, each controlling the inheritance of a particular trait. A gene is a segment of a chromosome that produces a particular trait. For example, in pea plants, there's a gene on the chromosome that codes for seed coat. The position of a gene on a chromosome is called a **locus**.

- A gene usually consists of a *pair* of hereditary factors. These factors are called **alleles**. Each organism usually carries two alleles for a particular trait. That is, alleles make up a gene, which in turn produce a trait. Alleles are alternate forms of the same gene. For example, if we're talking about the height of a pea plant, there's an allele for tall and an allele for short. In other words, both alleles are alternate forms of the gene for height.

- An allele can be **dominant** or **recessive**. In simple cases, an organism can express contrasting conditions. For example, a plant can be tall or short. The convention is to assign one of two letters for the two different alleles. The dominant allele receives a capital letter and the recessive allele receives a lowercase of the same letter. For instance, we might give the dominant allele for height in pea plants a "T" for tall. This means that the recessive allele would be "t."

- When an organism has two identical alleles for a given trait, the organism is **homozygous**. For instance, TT and tt both represent homozygous organisms. TT is homozygous dominant and tt is homozygous recessive. If an organism has two *different* alleles for a given trait, the organism is **heterozygous**.

- When discussing the physical appearance of an organism, we refer to its **phenotype**. The phenotype tells us what the organism looks like. When talking about the genetic makeup of an organism, we refer to its **genotype**. The genotype tells us which alleles the organism possesses.

One of the major ways ETS likes to test genetic information is by having you do crosses. Crosses involve the mating of hypothetical organisms with specific phenotypes and genotypes.

- Label each generation in the cross. The first generation in an experiment is always called the **parent**, or **P1 generation**. The offspring of the P1 generation are called the **filial**, or **F1 generation**. The next generation, the grandchildren, is called the **F2 generation**.

- Always write down the symbols you're using for the alleles, along with a key to remind yourself what the symbols refer to. Use uppercase for dominant alleles and lowercase for recessive alleles.

Now let's look at some basic genetic principles.

MENDELIAN GENETICS

One of Mendel's hobbies was to study the effects of cross-breeding on different strains of pea plants. Mendel worked exclusively with true-breeder pea plants. This means the plants he used were genetically pure and consistently produced the same traits. For example, tall plants always produced tall plants; short plants always produced short plants. Through his work he came up with three principles of genetics: the **law of dominance**, the **law of segregation**, and the **law of independent assortment**.

THE LAW OF DOMINANCE

Mendel crossed two true-breeding plants with contrasting traits: tall pea plants and short pea plants. This type of cross is called a **monohybrid cross**, which means that we're looking at only one trait. In this case, the trait is height.

To his surprise, when Mendel mated these plants, the characteristics didn't blend to produce plants of average height. Instead, all the offspring were tall.

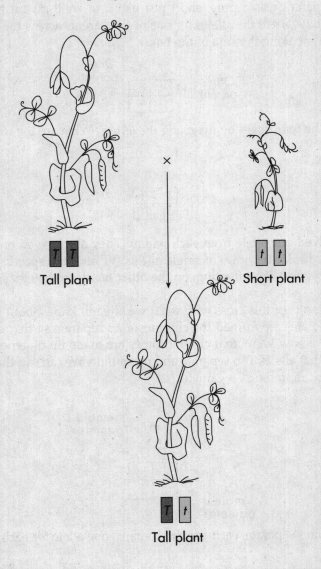

Mendel recognized that one trait must be masking the effect of the other trait. This is called the **law of dominance**. The dominant, tall allele, T, somehow masked the presence of the recessive, short allele, t. Consequently, all a plant needs is one tall allele to make it tall.

MONOHYBRID CROSS

A simple way to represent a cross is to set up a **Punnett square**. Punnett squares are used to predict the results of a cross. Let's construct a Punnett square for the cross between Mendel's tall and short pea plants. Let's first designate the alleles for each plant. As we saw earlier, we can use the letter "T" for the tall, dominant allele and "t" for the short, recessive allele.

Since one parent was a pure, tall pea plant, we'll give it two dominant alleles (TT homozygous dominant). The other parent was a pure, short pea plant, so we'll give it two recessive alleles (tt homozygous recessive). Let's put the alleles for one of the parents across the top of the box, and the alleles for the other parent along the side of the box.

```
                    T   T ← One
                          parent
    Other  → t  ┌────┬────┐
    parent      │    │    │
            t   ├────┼────┤
                │    │    │
                └────┴────┘
```

Now we can fill in the four boxes by matching the letters. What are the results for the F1 generation?

```
              T    T
          ┌────┬────┐
      t   │ Tt │ Tt │
          ├────┼────┤
      t   │ Tt │ Tt │
          └────┴────┘
```

Each offspring received one allele from each parent. They all received one T and one t. They're all Tt! Our parents had duplicate copies of single alleles, TT and tt, respectively. We could therefore refer to them as homozygous. The offspring, on the other hand, are heterozygous: they possess one copy of each allele.

Let's compare the results of this cross with what we already know about meiosis. From meiosis, we know that when gametes are formed the chromosomes separate so that each cell gets one copy of each chromosome. We now know that chromosomes are made up of genes, and genes consist of alleles. We've just seen that alleles also separate and recombine. We can say, therefore, that each allele in a Punnett square also "represents" a gamete:

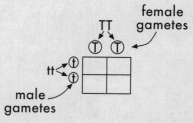

When the chromosomes separate, each gamete contains one allele for each trait.

THE LAW OF SEGREGATION

Next, Mendel took the offspring and self-pollinated them. Let's use a Punnett square to spell out the results. This time we're starting with the offspring of the first generation—F1. Take a look at the results:

	T	t
T	TT	Tt
t	Tt	tt

F2 Generation

One of the offspring was a short pea plant! The short-stemmed trait reappeared in the F2 generation. How did that happen? Once again, the alleles separated and recombined, producing a new combination for this offspring. The cross resulted in one offspring with a pair of recessive alleles, tt. Since there was no T (dominant) allele around to mask the expression of the short, recessive allele, our new plant wound up short.

Although all of the F1 plants appeared to be tall, the alleles separated and recombined during the cross. This is an example of the **law of segregation**.

What about the genotype and phenotype for this cross? Remember, genotype refers to the genetic makeup of an organism whereas phenotype refers to the appearance of the organism. Using the results of our Punnett square, what is the ratio of phenotypes and genotypes in the offspring?

Let's sum up the results. We have four offspring with two different phenotypes: three of the offspring are tall, whereas one of them is short. On the other hand, we have three genotypes: 1 TT, 2 Tt, and 1 tt.

Here's a summary of the results:

- The ratio of phenotypes is 3:1 (three tall: one short).

- The ratio of genotypes is 1:2:1 (one TT: two Tt: one tt).

THE LAW OF INDEPENDENT ASSORTMENT

So far, we have looked at only one trait: tall versus short. What happens when we study two traits at the same time? The two traits also segregate randomly. This is an example of **independent assortment**. For example, let's look at two traits in pea plants: height and color. When it comes to height, a pea plant can be either tall or short. As for color, the plant can be either green or yellow, with green being dominant. This gives us four alleles. By the law of independent assortment, these four alleles can combine to give us four different gametes:

TG Tg tG tg

Dihybrid Cross

Keep in mind that the uppercase letter refers to the dominant allele. Therefore, "T" refers to tall and "G" refers to green, whereas "t" refers to short and "g" refers to yellow. Now let's set up a cross between plants differing in two characteristics—called a **dihybrid cross**—using these four gametes and see what happens.

Each trait will act independently, meaning that a plant that is tall can be either green or yellow. Similarly, a green plant can be either tall or short:

Here is the Punnett Square for a cross between two double heterozygotes (Tt Gg):

	TG	Tg	tG	tg
TG	TG TG	Tg TG	tG TG	tg TG
Tg	TG Tg	Tg Tg	tG Tg	tg Tg
tG	TG tG	Tg tG	tG tG	tg tG
tg	TG tg	Tg tg	tG tg	tg tg

This is an example of the law of independent assortment. Each of the traits segregated independently. Don't worry about the different combinations in the cross—you'll make yourself dizzy with all those letters. Simply memorize the phenotype ratio of the pea plants. For the 16 offspring there are:

- 9 tall and green
- 3 tall and yellow
- 3 short and green
- 1 short and yellow

That's 9:3:3:1. Since Mendel's laws hold true for most of the traits they'll ask you about on the AP test, simply learning the ratios of offspring for this type of cross will help you nail any questions that come up.

The Punnett square method works well for monohybrid crosses and helps us visualize the possible combinations. However, a better method for predicting the likelihood of certain results from a dihybrid cross is to apply the law of probability. For dihybrid ratios, the law states that the probability that two or more independent events will occur simultaneously is equal to the *product* of the probability that each will occur independently. To illustrate the product rule, let's consider again the cross between two identical dihybrid tall, green plants with the genotype TtGg. To find the probability of having a tall, yellow plant simply *multiply* the probabilities of each event. If the probability of being tall is $\frac{3}{4}$ and the probability of being yellow is $\frac{1}{4}$, then the probability of being tall *and* yellow is $\frac{3}{4} \times \frac{1}{4} = \frac{3}{16}$.

One more thing: The probability can be expressed as a fraction, percentage, or a decimal. Remember that this rule works only if the results of one toss are not affected by the results of another toss.

Let's summarize Mendel's three laws.

SUMMARY OF MENDEL'S LAWS	
Laws	**Definition**
Law of Dominance	One trait masks the effects of another trait.
Law of Segregation	Alleles can segregate and recombine.
Law of Independent Assortment	Traits can segregate and recombine independently of other traits.

TEST CROSS

Suppose we want to know if a tall plant is homozygous (TT) or heterozygous (Tt). Its physical appearance doesn't necessarily tell us about its genetic makeup. The only way to determine its genotype is to cross the plant with a recessive, short plant, tt. This is known as a **test cross**. Using the recessive plant, there are only two possibilities: (1) TT × tt or (2) Tt × tt. Let's take a look.

If none of the offspring is short, our original plant must have been homozygous, TT. If, however, even one short plant appears in the bunch, we know that our original pea plant was heterozygous, Tt. In other words, it wasn't a pure-breeding plant. A test cross uses a recessive organism to determine the genotype of an organism of unknown genotype.

BEYOND MENDELIAN GENETICS

Not all patterns of inheritance obey the principles of Mendelian genetics. In fact, many traits we observe are due to a combined expression of alleles. Here are a couple of examples of non-Mendelian forms of inheritance:

- **Incomplete dominance:** In some cases, the traits will blend. For example, if you cross a white snapdragon plant (dominant) and a red snapdragon plant (recessive), the resulting progeny will be pink.

- **Codominance:** Sometimes you'll see an equal expression of both alleles. For example, an individual can have an AB blood type. In this case, each allele is equally expressed. That is, both the A allele and the B allele are expressed (I^AI^B). That's why the person is said to have AB blood.

- **Polygenetic inheritance:** In some cases, a trait results from the interaction of *many* genes. Each gene will have a small effect on a particular trait. Height, skin color, and weight are all examples of polygenetic traits.

- **Multiple alleles**: Some traits are the product of many different alleles that occupy a specific gene locus. The best example is the ABO blood group system in which three alleles (I^A, I^B, and i) determine blood type.

- **Epistasis**: In some cases, the genes at one locus may influence the expression of genes at another locus. For example, two gene loci affect the coat color of mice. In one case, black (B) is dominant to brown (b). Yet, at another gene locus a pair of alleles (C) and (c) also affect coat color. When an albino mouse from a true-breeding white strain (cc) and a mouse from a true-breeding brown strain (bb) reproduce the offspring are all black (CcBb). In this example, the recessive albino genotype is epistatic to the brown/black genotype.

- **Pleiotrophy:** Sometimes an allele can affect a number of characteristics of an organism. For example, in sickle cell anemia, multiple symptoms (such as pneumonia, heart failure, and limited mental functioning) are caused by a single pair of alleles.

- **Linked genes**: Sometimes genes on the *same* chromosome stay together during assortment and move as a group. The group of genes is considered linked and tends to be inherited together. For example, the genes for flower color and pollen shape are linked on the same chromosomes and show up together. This pattern has led to methods for mapping human chromosomes.

One more thing: Since linked genes are found on the same chromosome, they cannot segregate independently. That means they do not follow the probability rule and the expected results from a dihybrid cross.

Linked genes stay together and do not assort independently. Scientists also discovered that the frequency of crossing-over between any two linked alleles was proportional to the distance between them. That is, the further apart two linked alleles are on a chromosome, the more often the chromosome will break between them. This finding led to recombination mapping—mapping of linkage groups with each map unit being equal to 1 percent recombination. For example, if two linked genes, A and B, recombine with a frequency of 15 percent, and B and C recombine with a frequency of 8%, and A and C recombine with a frequency of 24 percent, what is the sequence and the distance between them?

The sequence and the distance of A-B-C is:

```
        15 units              9 units
A_____B_____C
```

If the recombination frequency between A and C had been 6 percent instead of 24 percent, the sequence and distance of A-B-C would now be:

```
      6 units        9 units
A_____C_____B
```

SEX-LINKED TRAITS

We already know that humans contain 23 pairs of chromosomes. Twenty-two of the pairs of chromosomes are called **autosomes**. They code for many different traits. The other pair contains the **sex chromosomes**. This pair determines the sex of an individual. A female has two X chromosomes. A male has one X and one Y chromosome—an X from his mother and a Y from his father. Some traits, such as **color blindness** and **hemophilia**, are carried on sex chromosomes. These are called **sex-linked traits**. Most sex-linked traits are found on the X chromosome and are more properly referred to as "X-linked."

Since males have one X and one Y chromosome, what happens if a male has a defective X chromosome? Unfortunately, he'll express the sex-linked trait. Why? Because his one and only X chromosome is defective. He doesn't have another X to mask the effect of the bad X. However, if a female has only one defective X chromosome, she won't express the sex-linked trait. For her to express the trait, she has to inherit two defective X chromosomes. A female with one defective X is called a **carrier**. But she's not completely safe. Although she *appears* normal, she can still pass on the trait to her children. If, however, she inherits two copies of the defective allele, she's out of luck. She will express the trait.

You can also use the Punnett square to figure out the results of sex-linked traits. Here's a classic example: A male who has normal color vision and a woman who is a carrier for color blindness have children. How many of the children will be color-blind? To figure out the answer, let's set up a Punnett square:

$$
\begin{array}{c c}
 & \text{X} \quad \bar{\text{X}} \; \leftarrow \text{Mother} \\
\text{Father} \rightarrow \begin{array}{c} \text{X} \\ \text{Y} \end{array} & \begin{array}{|c|c|} \hline \text{XX} & \overline{\text{XX}} \\ \hline \text{XY} & \overline{\text{XY}} \\ \hline \end{array}
\end{array}
$$

$$\bar{\text{X}} = \text{diseased X}$$

Notice that we placed a bar above any defective X to indicate the presence of a defective allele. And now for the results. The couple would have one son who is color-blind, a normal son, a daughter who is a carrier, and a normal daughter. The sick child happens to be a son.

Barr Bodies

A look at the cell nucleus of normal females will reveal a dark-staining body known as a **Barr body**. A Barr body is an X chromosome that is condensed and visible. In every female cell, one X chromosome is activated and the other X chromosome is deactivated during embryonic development. Surprisingly, the X chromosome destined to be inactivated is randomly chosen in each cell. Therefore, in every tissue in the adult female one X chromosome remains condensed and inactive. However, this X chromosome is replicated and passed on to a daughter cell.

KEY WORDS

Gregor Mendel
trait
genes
locus
alleles
dominant
recessive
homozygous
heterozygous
phenotype
genotype
parent, or P1 generation
filial, or F1 generation
F2 generation
monohybrid cross
Punnett square
law of dominance

law of segregation
law of independent assortment
dihybrid cross
test cross
incomplete dominance
codominance
polygenetic inheritance
multiple alleles
epistasis
pleiotrophy
linked genes
autosomes
sex chromosomes
hemophilia
color blindness
sex-linked traits
carrier

QUIZ

<u>Directions</u>: Each of the questions or incomplete statements below is followed by five suggested answers or completions. Select the one that is best in each case. Answers can be found on page 241.

1. In pea plants, the allele for smooth seeds (S) is dominant over the allele for wrinkled seeds (s). In an experiment, when two hybrids are crossed, what percent of the offspring share the same genotype as the parents?

   (A)   0%
   (B)   25%
   (C)   50%
   (D)   75%
   (E)   100%

2. An organism has three independently assorting traits AaBbCc. What fraction of its gametes will contain the recessive genes abc?

   (A)   0

   (B) $\dfrac{1}{8}$

   (C) $\dfrac{1}{4}$

   (D) $\dfrac{1}{2}$

   (E) $\dfrac{3}{4}$

3. Which of the following findings provides the best evidence that an abnormal trait is sex-linked?

   (A)   The trait always skips a generation.
   (B)   Some members of a family are carriers of the disease.
   (C)   The trait appears in all of the offspring.
   (D)   The trait is passed from mother to daughters.
   (E)   The trait is passed from mother to sons.

4. In the fruit fly *Drosophila*, the allele for normal body (B) is dominant to the allele for hairy body (b). When two normal-bodied fruit flies were mated they produced 81 hair-bodied flies and 319 normal-bodied flies. The genotypes of the parents are most likely

   (A)   BB × bb
   (B)   BB × Bb
   (C)   Bb × Bb
   (D)   Bb × bb
   (E)   bb × bb

5. If a coin is tossed three consecutive times, what is the probability of getting three heads?

(A) $\dfrac{1}{8}$

(B) $\dfrac{1}{6}$

(C) $\dfrac{1}{4}$

(D) $\dfrac{1}{2}$

(E) $\dfrac{3}{4}$

<u>Directions</u>: Each group of questions consists of five lettered headings followed by a list of numbered phrases or sentences. For each numbered phrase or sentence, select the one heading that is most closely related to it and fill in the corresponding oval on the answer sheet. Each heading may be used once, more than once, or not at all in each group.

<u>Questions 6–9</u>

(A) Phenotype
(B) Codominance
(C) Heterozygous
(D) Dihybrid cross
(E) Monohybrid cross

6. Alleles for two different traits are observed in a single organism

7. The functional manifestations of gene activity in an organism

8. The individual expression of two inherited alleles

9. A cross involving two unlinked pairs of alleles

9
Classification

TAXONOMY

Man has long been aware of the diversity of life. But figuring out exactly how organisms are related to one another took some work. Early scientists agreed that many creatures shared common features. However, there was little agreement about how all organisms were related to one another, if at all!

Interestingly, our means of classifying organisms is essentially the same as that used by those earliest scientists: We order organisms into groups on the basis of shared characteristics or traits. The major difference between our approach and that of early biologists is that we now know that such traits are acquired over the long process of evolution. Consequently, we group animals together on the basis of evolutionary relatedness, and not simple appearance. These relations are manifested in traits, and the sum of these traits is known as **phylogeny**. The science of classifying animals according to their phylogeny is known as **taxonomy**.

HOW CLASSIFICATION WORKS

The order of classification from fewest to most characteristics in common is:

Kingdom
Phylum
Class
Order
Family
Genus
Species

Notice that as we move down the list, we find that organisms have more and more in common. Consequently, each step down the scale includes fewer and fewer members. The kingdom is the biggest group, and includes the greatest number of members, while the species is the smallest group, and counts the fewest members. Moreover, organisms in the same phylum have more in common than those in the same kingdom. Those in the same class have more in common than those in the same phylum, and so on. This means that as we move from top to bottom, we go from "less in common" to "more in common."

NAMING NAMES

The **binomial classification** system was developed by **Carolus Linnaeus**. By "binomial," we simply mean that all organisms are given scientific names consisting of a species name and a genus name, usually in Latin or in a Latinate version of the discoverer's name. For example, we humans are called *Homo sapiens*. *Homo* is our genus name and *sapiens* is our species name. *Homo* means "man" and sapiens means "wise."

For the AP Biology Exam, you're supposed to memorize the order of classification. Do you remember the mnemonic we saw in Chapter 1?

King Philip of Germany decided to walk to America. What do you think happened?

King	-	Kingdom
Philip	-	Phylum
Came	-	Class
Over	-	Order
From	-	Family
Germany	-	Genus
Soaked	-	Species

THE NITTY GRITTY OF CLASSIFICATION

Now that we've discussed the classification scheme, let's take a closer look at how organisms are actually ordered into various levels. We'll start with the highest level, the kingdoms. All organisms belong to one of five kingdoms: **Monera**, **Protista**, **Fungi**, **Plantae**, and **Animalia**. Here's a quick summary of the kingdoms along with some of the principal characteristics you'll need to know for the AP Test.

I. KINGDOM MONERA

Monerans are **prokaryotes**. They are one-celled organisms that lack a nucleus and membrane-bound organelles. Monerans have a single, circular DNA molecule that does not contain the histones found in eukaryotic chromosomes. These prokaryotes are classified by their size, shape, habitat, metabolism and energy sources. The kingdom includes eubacteria and archaebacteria.

A. Archaebacteria

Archaebacteria are primitive organisms that have unique characteristics. Their cell wall lacks peptidoglycan found in eubacteria and their plasma membranes contain unusual fatty acids. Another difference is that their ribosomes look more like those of eukaryotes. Some archaebacteria derive energy by converting CO_2 and H_2 to methane (**methanogenic**) while others use H_2S as their source of energy (**chemoautotrophs**). They include bacteria that live in hot, acid environments (**thermophilic**) and those that only live in salty ponds (**halophiles**).

B. Eubacteria

The largest class of Monerans is eubacteria. They possess a cell wall made of peptidoglycan. Eubacteria are extremely diverse, especially in the way they obtain nutrients. Some are chemoautotrophs (**nitrifying bacteria**), some are photosynthetic autotrophs (**cyanobacteria**), but most are heterotrophs. Within the heterotrophs, some are **decomposers** (breaking down organic material) and others are **pathogens** (disease-causing parasites). Some eubacteria have a flagellum.

Here's a simple illustration of your basic bacterium:

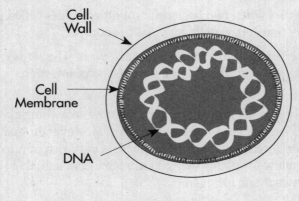

Bacterium

Cell Wall

The composition of bacterial cell walls isn't that important for us. However, ETS might want you to know that peptidoglycan enables us to identify certain types of bacteria through "Gram staining." A bacteria with a thick peptidoglycan cell wall will test Gram-positive, whereas one with a thin peptidoglycan layer will test Gram-negative.

Bacteria also vary in size and shape. They can be **cocci** (spherical), **bacilli** (rod-shaped), and **spirilli** (corkscrewed). Some bacteria also have flagella.

| Bacilli | Cocci | Spirilli |

Oxygen Use

As far as respiration is concerned, most bacteria need oxygen. We call these bacteria **obligate aerobes**. Yet other bacteria cannot survive in the presence of oxygen. These bacteria are known as **obligate anaerobes** and live in places like the deep-sea floor, where sulfur-rich vents open up from the earth's core. Still other bacteria are known as **facultative anaerobes**, which means that they can use oxygen if it's available, and won't if it isn't available.

Reproduction

Bacteria reproduce asexually by binary fission. Binary fission occurs when the bacteria replicate their chromosomes and divide into two identical cells. While bacteria are not *sexual*, they do exchange genetic material. This exchange of genetic information is known as **genetic recombination** and happens in one of three ways: **transformation**, **conjugation**, or **transduction**.

1. **Transformation** occurs when a bacterium picks up naked DNA from the environment.

2. **Conjugation** occurs when two bacteria form little bridges called **pili** between one another and one transfers genetic material to another.

3. **Transduction** occurs with the intervention of a virus. The virus carries some DNA from one bacterium to another during the process of infection.

There are a few special types of bacteria you should know something about. One of them is called **nitrogen-fixing bacteria**. Plants need nitrogen to survive. However, in many places, the soil is relatively poor in nitrogen. You'll recall that the atmosphere is very rich in nitrogen (some 78 percent of the air is nitrogen). Fortunately for plants living in nitrogen-poor soil, there are bacteria that are able to utilize, or "fix", this atmospheric nitrogen.

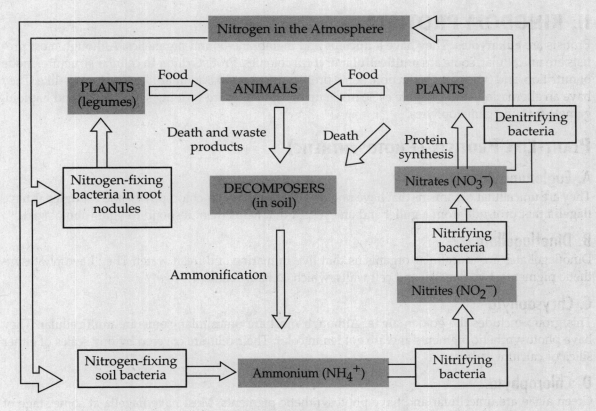

The Nitrogen Cycle

The plants that have this special relationship with nitrogen-fixing bacteria are known as **legumes**. Pea plants and clover plants are examples of legumes. The nitrogen-fixing bacteria set up house in the root nodules of these plants, forming a mutualistic relationship: Both organisms are happy. The plants get their nitrogen and the bacteria get their shelter. We'll talk more about symbiotic relationships a little later on. Nitrogen-fixing bacteria have symbiotic relationships with certain plants.

Now that we've given you an overview of the classification of organisms, let's move on to protista.

KEY WORDS

phylogeny
taxonomy
kingdom
phylum
class
order
family
genus

species
binomial classification
Carolus Linnaeus
monera
protista
fungi
plantae
animalia

II. KINGDOM PROTISTA

Protists are **eukaryotes**. They have a **nucleus** and membrane-bound organelles. Although most protists are unicellular, some are **multicellular** or form colonies. Protists differ in cellular structure, mode of nutrition, and type of reproduction. Some protists are plant-like, animal-like, and fungus-like. They have an alternating, two-part, life cycle made up of diploid, spore-forming sporophytes and haploid, gamete-forming gametophytes.

PLANT-LIKE PROTISTS (PHOTOSYNTHETIC)

A. Euglenophyta

They are unicellular organisms that have photosynthetic pigments, chlorophyll *a* and *b*. **Euglenas** have **flagella** that protrude from a gullet and an eyespot that helps them respond to light (phototaxis).

B. Dinoflagellata

Dinoflagellates are unicellular organisms that live in marine and fresh water. They have photosynthetic pigments, two flagella, and cell walls, which contain cellulose.

C. Chrysophyta

This group includes the golden **algae**. Although most are unicellular, some are multicellular. They have photosynthetic pigments and are golden in color. Their cells are covered by tiny scales of either silica or calcium carbonate.

D. Chlorophyta

Green algae are unicellular and have photosynthetic pigments. Most have flagella at some stage of their life and store food as starch.

E. Phaeophyta

Brown algae are mostly multicellular and have two sperms with flagella on reproductive cells. They are commonly known as "seaweeds."

F. Rhodophyta

Red algae are mostly multicellular, marine seaweeds. In addition to chlorophyll *a* and *b*, they have red photosynthetic pigments (phycobilins).

G. Bacillariophyta

Diatoms are usually nonmotile, unicellular organisms with cell walls made of silica.

ANIMAL-LIKE PROTISTS (NONPHOTOSYNTHETIC HETEROTROPHS)

H. Zoomastigina

The zooflagellates are unicellular protozoans that move by means of a flagellum. Some live in the gut of termites (*Trichonympha*); others are parasitic and cause disease such as African sleeping sickness (*Trypanosoma*).

I. Rhizopoda

Amoebas perform **phagocytosis** by surrounding and engulfing food using pseudopods ("false feet").

J. Ciliophora

They are unicellular protozoans, which move around and feed using their tiny hairs (**cilia**). The best known is the **paramecium**. They possess two nuclei, a mouth, and contractile vacuoles.

K. Sporozoa

Sporozoans are nonmotile, parasitic spore-formers. They are characterized by their lack of flagella and an amoeboid body form. They include the *Plasmodium*, which causes malaria.

L. Foraminifera

These unicellular protists produce calcareous tests (shells) with pores through which cytoplasmic projections extend.

FUNGUS-LIKE PROTISTS

M. Myxomycota

Slime molds produce large multinucleated masses (plasmodium) Sometimes slime molds have stalks, which grow upward and spores form (fruiting bodies). Other times, they produce **gametes**, which fuse and produce a diploid zygote to form a multinucleated mass. They are found in moist soil, decaying leaves, or logs in a damp forest.

III. KINGDOM FUNGI

A. Fungi

Fungi are generally **multicellular** organisms with cell walls made of chitin. Others like yeast are unicellular. Since they lack chlorophyll they are heterotrophs. They feed using threadlike branches (called *hyphae*) that secrete digestive enzymes into food and absorb the products. The group includes parasites, pathogens (athlete's foot), and decomposers. Fungi reproduce either by forming **spores** or by **budding**.

IV. KINGDOM PLANTAE

Plants are multicellular, photosynthetic organisms with cell walls made of cellulose. In plants, the fertilized egg develops in a multicellular embryo within a protective archegonium. They reproduce sexually and asexually, with alternating gametaphyte and sporophyte generations.

A. Bryophytes

Bryophytes are the more primitive plants and are characterized by the lack of true stems, roots, and leaves. Bryophytes anchor themselves in the soil by **rhizoids** (root-like structures). Bryophytes have flagellated sperms within the **antheridium** (male-containing part) that swim to reach the egg, which lie within the **archegonium** (the female gametophyte). Common bryophytes are mosses, liverworts, and hornworts.

B. Pterophyta

Ferns are the "seedless plants." They are among the earliest vascular plants to colonize land. The life cycle of ferns involve alternation of generation, in which the dominant stage is the sporophyte generation.

C. Spenophyta

These vascular plants with hollow, ribbed stems and reduced, scalelike leaves are the horsetails. Although the extinct species were once as large as modern trees, the surviving species are small and found in wet, marshy habitats.

D. Lycophyta

Club mosses are small plants with rhizomes and short, erect branches. Like horsetails, they were common 300 million years ago when the extinct species were treelike plants.

E. Coniferophyta

Conifers are the woody plants that bear their seeds in cones (the seeds are not enclosed). They have tracheids and well-developed phloem. Both the roots and stems are capable of secondary growth. Fertilization doesn't require a water source.

F. Anthophyta

These are known as the "flowering plants." They have seeds enclosed within a fruit or nut. They are multicellular plants with highly specialized and developed conducting tissue for the transport of water and nutrients. Angiosperms can be further divided into two classes: monocots and dicots.

    a) **Monocots**
       Monocots have a single **cotyledon** (the embryonic seed leaf). The characteristics include flower parts in threes or multiples of threes, vascular tissues usually in scattered bundles, a fibrous root system, and leaves with parallel veins.

    b) **Dicots**
       Dicots have two **cotyledons**. The characteristics include flower parts mostly in fours or fives, vascular tissue in distinct bundles arranged in a circle, a taproot system, and leaves with netted veins.

V. KINGDOM ANIMALIA

Animals are multicellular, heterotrophic organisms that are extremely diverse. For the AP Biology Exam, some of the questions will focus on the following characteristics: body symmetry, tissue complexity, type of body cavity, and developmental patterns, and how they were major milestones of animal evolution. For instance, there are two basic types of bodies: **radial symmetry** (circular or disk-shaped form) and **bilateral symmetry** (a cut through the midline divides the organism into mirror images). Animals then evolved from two cell layers to three layers.

With the development of bilateral symmetry, animals developed a body cavity. The early animals lacked a body cavity (**acoeloms**). Then some developed a body cavity lined with tissue not completely derived from mesoderm (**pseudocoeloms**), while others developed a true body cavity lined with mesoderm (coeloms). Another fundamental split among animals is when the mouth and anus form. **Protostomes** undergo spiral cleavage and the mouth forms first. **Deuterostomes** undergo radial cleavage and the anus forms first.

A. Porifera

Sponges are sessile (nonmotile) organisms whose bodies are constructed of two cell layers. They ingest food by drawing a steady current of water through their pores. Sponges have radial symmetry.

B. Cnidaria

Cnidarians have a two-cell layered body with a digestive cavity surrounded by tentacles that sting their prey. Common cnidarians are jellyfish, hydras, and sea anemones.

C. Platyhelminthes

Flatworms are motile organisms whose bodies are the first to have three cell layers and bilateral symmetry. They are also acoelomates (no body cavity). Platyhelminthes can be parasitic, and can undergo regeneration.

D. Nematoda

Roundworms are soil-dwellers that have pseudocoelomate bodies. They have a complete digestive tract that extends from mouth to anus. Some are parasites.

E. Rotifera

Rotifers are tiny filter feeders that are pseudocoelomates with a complex, complete digestive system.

F. Mollusca

Mollusks are motile organisms with soft bodies and hard shells. They are the first protostomes and the first coelomates (with spiral, determinate cleavage). Members of the mollusks include octopuses, squids, snails and clams.

G. Annelida

Annelids are segmented worms with two openings: a mouth and anus. They have a fully developed digestive system, a closed circulatory system, a developed nervous system, and bristle appendages (setae). The most common examples are earthworms.

H. Arthropoda

Arthropods are animals with segmented bodies; paired, jointed legs; and a chitinous exoskeleton. They are unusual in that they have an open circulatory system with a dorsal heart. Examples include insects, arachnids, and crustaceans.

I. Echindermata

Echinoderms are sessile or sedentary animals with a spiny exoskeleton. They are the first deuterosomes (the blatospore develops into the anus). An example of an echindermata is a sea urchin.

J. Chordata

Chordates are animals that have a **notochord**, a **dorsal nerve cord**, a postanal tail, and **pharyngeal gill slits** at some time in their lives. Although most are vertebrates (have backbones), some are not. The invertebrates include tunicates, amphioxus, and the acorn worms.

i. **Pisces**
Fish are cold-blooded vertebrates that have gills, scales, and a two-chambered heart.

ii. **Amphibians**
Amphibians initially breathe through gills, and then develop lungs. They can also exchange gas through their moist skin. They have a three-chambered heart.

iii. **Reptilia**
Reptiles are cold-blooded animals that have eggs with a chitinous covering. They have a four-chambered heart and are the first vertebrates to have internal fertilization.

iv. **Aves**
Birds are warm-blooded, have eggs with shells, wings, feathers, hollow bones, and a four-chambered heart.

v. **Mammalia**
Mammals are warm-blooded animals with a four-chambered heart. They have hair and produce milk to feed their young. Some mammals have a placenta (structure that joins the embryo to the wall of the mother's uterus). Marsupials, on the other hand, do not have a placenta and the developing embryo receives very little nourishment from the mother in the uterus. About eight days after fertilization, the fetus must continue its maturation in its mother's pouch.

VIRUSES

Now let's talk about **viruses**. Viruses are the smallest organisms that we'll have to deal with on the AP Biology Exam. They might be tiny, but they have an enormous impact on many organisms, including humans: Such diseases as hepatitis, the common cold, and AIDS are all caused by viruses.

Scientists believe that viruses don't belong with the other organisms we've just discussed. This is because they are *not* true cells. In fact, they aren't really "alive" in the classic sense. When we say that a virus is not "alive," we mean that it doesn't live or reproduce independently.

Viruses have only two basic components:

- A protein coat made of proteins.

- A viral chromosome, which can be *either* **DNA** or **RNA**.

Because they will most likely appear on the AP Biology Exam, let's discuss how they reproduce:

1. A virus attaches to receptors on a "host" cell. This can be any type of cell—a bacterium, a plant, or an animal cell.

2. Once attached, the virus injects its nucleic acid *into* the host cell.

3. The virus's nucleic acid *uses the host's DNA* to synthesize the necessary components for its replication.

4. Newly synthesized viral components are assembled.

5. Eventually, viruses produced in the cell are released and lytic enzymes kill the host cell.

Take a look at the lytic cycle of a typical bacterial virus known as a bacteriophage:

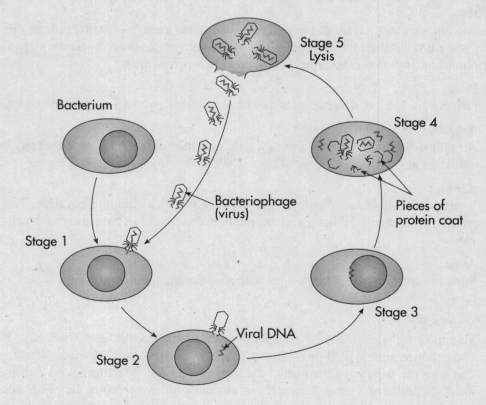

Lysogenic Cycle

Sometimes a virus can integrate its DNA into the host DNA, making it a **lysogenic cell**. In this case, when the bacterial DNA replicates, the viral DNA also replicates. The **temperate virus** will not destroy its host as long as the viral genes that code for viral structural proteins are repressed. However, when certain conditions cause a temperate virus to revert to a lytic cycle, it will then destroy the host cell.

Many scientists speculate that viruses are in fact simply "renegade" parts of the genetic code. In other words, during normal replication, something went wrong, sending out a piece of the code that, through mutations, came to be able to replicate itself, utilizing its original parent, or "host," machinery.

KEY WORDS

cyanobacteria
bacteria
cocci
bacilli
spirilli
obligate aerobes
obligate anaerobes
facultative anaerobes
genetic recombination
transformation
conjugation
transduction
pili
nitrogen-fixing bacteria
legumes
algae
protists
amoeba
phagocytosis

paramecium
cilia
euglena
flagellum
unicellular
multicellular
multinucleate
sporophytes
parasites
spores
gametes
alternation of generations
budding
viruses
DNA
RNA
lysis
lysogenic cell
temperate virus

QUIZ

<u>Directions:</u> Each of the questions or incomplete statements below is followed by five suggested answers or completions. Select the one that is best in each case. Answers can be found on page 242.

1. A unicellular organism is discovered that has the following characteristics: a nucleus, a cell wall, and photosynthetic organelles. This organism is most likely to be classified as

   (A) a protist
   (B) a fungi
   (C) a plant
   (D) an animal
   (E) a bacteria

2. Insects and spiders are classified as arthropods because they are both equipped with

(A) jointed legs, a chitinous exoskeleton, and wings
(B) segmented bodies, a chitinous exoskeleton, and wings
(C) segmented bodies, a chitinous exoskeleton, and a closed circulatory system
(D) jointed appendages, segmented bodies, and an open circulatory system
(E) jointed legs, three body parts, and a chitinous exoskeleton

3. A student observes an organism under a microscope. It is unicellular, eukaryotic, has threadlike branches, and a chitinous cell wall. Which kingdom is the organism most likely to be classified under?

(A) Fungi
(B) Monera
(C) Protista
(D) Animalia
(E) Plantae

4. Which of the following animals possess nephridia?

(A) Fish
(B) Amphibians
(C) Flatworms
(D) Earthworms
(E) Insects

5. All of the following are classes of vertebrates EXCEPT:

(A) Reptilia
(B) Arthropoda
(C) Mammalia
(D) Amphibia
(E) Aves

6. All of the following are true concerning fungi EXCEPT:

(A) they reproduce sexually and asexually.
(B) they are eukaryotic.
(C) they are photosynthetic.
(D) they require O_2.
(E) they have a cell wall.

7. Nitrogen-fixing bacteria are abundant in soils and convert atmospheric nitrogen into nitrates. Plants are able to use the nitrates to synthesize plant proteins. This is an example of

(A) mutualism
(B) parasitism
(C) commensalism
(D) competition
(E) tropism

8. Which of the following characteristics is (are) found in all viruses?

(A) A nuclear membrane
(B) A cell wall
(C) A protein coat
(D) Membrane-bound organelles
(E) Ribosomes

Directions: Each group of questions consists of five lettered headings followed by a list of numbered phrases or sentences. For each numbered phrase or sentence, select the one heading that is most closely related to it and fill in the corresponding oval on the answer sheet. Each heading may be used once, more than once, or not at all in each group.

Questions 9–12

(A) Conjugation
(B) Transduction
(C) Transformation
(D) Plasmid
(E) Spores

9. The transfer of genes between two *E. coli* joined by a sex pilus

10. A section of bacterial DNA is packaged in a virus and transferred to a new host cell

11. Small rings of bacterial DNA that carry accessory genes

12. A bacteria takes up segments of naked DNA

10

Plants

Let's begin our discussion of multicellular organisms with the plant kingdom. All told, the plant kingdom includes thousands of species, which have established themselves in every possible habitat. With such a wide range of habitats, plants have naturally taken on a dazzling variety of forms: There are over 260,000 different species of flowering plants alone!

ETS expects you to be familiar with the general characteristics, life cycles, and classifications of plants. We've already covered some of these characteristics in earlier chapters. You should recall that plants:

- Are multicellular, eukaryotic organisms

- Have a cell wall made of cellulose

- Are photosynthetic, meaning that they convert light energy to chemical energy by means of chloroplasts located primarily in their leaves

- Take up water via capillary action

PLANT CLASSIFICATION

Since we've already discussed plant classification, we're going to review the principal subdivisions. Plant classification is also useful because it gives you an idea of the evolutionary history of plants.

Let's start with a simple flow chart of the different subdivisions of plants:

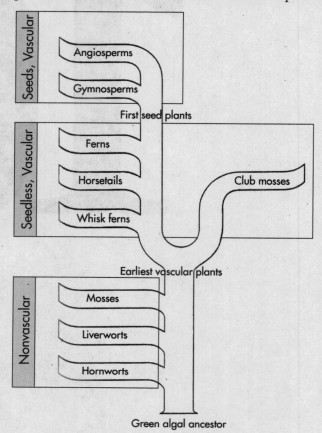

You'll notice from our chart that all plants fall into three major groups: **nonvascular**, **seedless vascular**, and **seeded vascular**. Bryophytes are the more primitive plants: They lack true stems, roots, and leaves. Tracheophytes are more advanced plants that possess specialized conducting and **vascular tissues**. As we'll see later on, these tissues are essential for the transport of material throughout the plant.

BRYOPHYTES VERSUS TRACHEOPHYTES

Bryophytes are the simplest plants, and are characterized by their lack of true stems and leaves. Common bryophytes include **mosses** and **liverworts**. They lack the vascular tissues—stems, roots, and leaves. The lack of specialized transport tissues is a major limitation for these simple plants.

The other phylum, tracheophyta, includes the most common and widespread of land plants. Plants belonging to this phylum have true vascular tissues. These tissues enable them to thrive on land by facilitating the transport and storage of water and nutrients. Since vascular tissues are the deciding factor in splitting plants into the two major phyla, let's take a closer look at them.

Vascular Tissues

Tracheophytes contain two types of vascular tissues: **xylem** and **phloem**. Xylem is tissue that conducts water and minerals up a plant from its roots. There are two types of xylem cells: **tracheids** and **vessel elements**. The tracheids are long and thin; the vessel elements are short and thick.

Water enters the plant through the **roots**. Roots have special features in their outer layer called **root hairs** that increase the surface area for absorption of materials.

In addition to absorbing water, minerals, and nutrients, roots also anchor the plant in the soil.

Phloem vessels carry nutrients, such as glucose, throughout the plant. Phloem cells are made up of **sieve tube elements** and **companion cells**. Sieve tube elements are the cells that actually carry the nutrients in a plant. The companion cells hang around to lend "support" to sieve tube elements. Ferns, trees, and flowering plants are all examples of tracheophytes.

Root hairs

Let's recap:

- Xylem carry water and minerals.

- Phloem carry nutrients.

- Root hairs carry water and increase the surface area for absorption.

- Roots also serve to anchor the plant.

Of all tracheophytes, ferns are the simplest and the most ancient. Enormous ferns covered much of the earth even before the dinosaurs appeared. Ferns are known as the "seedless plants." They're able to transport water, minerals, and nutrients throughout the plant because they have vascular tissue. However, since they're seedless, they still need an abundant water supply for fertilization. As we saw above, water is necessary for all land-based organisms, even ferns, in order for the sperm to make their journey to the egg cell in reproduction.

GYMNOSPERMS AND ANGIOSPERMS: THE SEEDED PLANTS

Gymnosperms

Tracheophytes can be further subdivided into **gymnosperms** and **angiosperms**. These are the "seeded plants."

Gymnosperms are among the woody plants and are, evolutionarily speaking, among the oldest plants around. They include such common evergreens as spruces, hemlocks, and firs and are **perennial**, meaning they live year after year. One way to determine the age of a gymnosperm is by counting the number of **tree rings**. These rings, which are composed of dead xylem, represent the tree's annual growth.

Angiosperms

Of all the plants on the planet, the angiosperms are the most varied and widespread. Angiosperms are also known as the "**flowering plants**." They have enclosed seeds located within a fruit or nut. Some flowering plants are woody such as oak, cherry, and walnut. Angiosperms can be further divided into two classes: **monocots** and **dicots**. The monocots have a single **cotyledon**—the embryonic seed leaf. Monocots are known to have leaves with parallel veins and flower parts in multiples of three. Some examples are orchids and lilies. Dicots, on the other hand, have two cotyledons. They have broad leaves with netted veins, and flower parts in multiples of four or five. Take a look at the major differences between monocots and dicots.

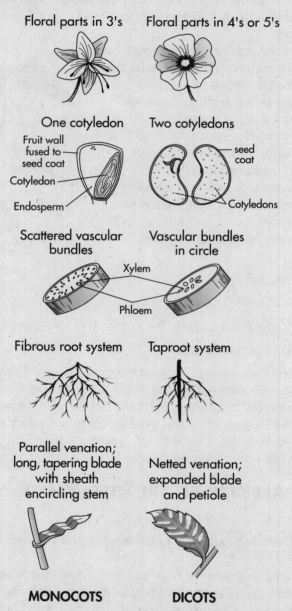

Floral parts in 3's   Floral parts in 4's or 5's

One cotyledon   Two cotyledons

Fruit wall fused to seed coat
Cotyledon
Endosperm
seed coat
Cotyledons

Scattered vascular bundles   Vascular bundles in circle
Xylem
Phloem

Fibrous root system   Taproot system

Parallel venation; long, tapering blade with sheath encircling stem   Netted venation; expanded blade and petiole

MONOCOTS   **DICOTS**

LIFE CYCLE OF PLANTS

Now that we've looked at the differences in plant classification, we're ready to review the life cycles and reproduction of plants. One thing you'll need to know for the AP Biology Exam is that plants have an **alternation of generation**. That is, they spend part of their lives as haploids and part of their lives as diploids.

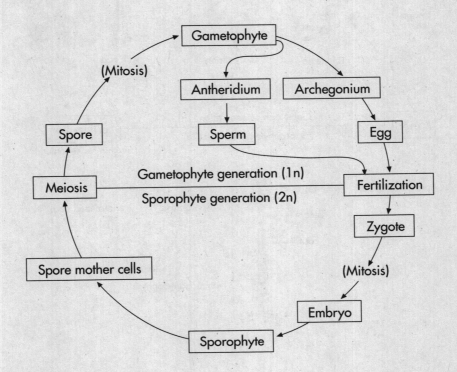

The haploid plant is called a **gametophyte** because it produces haploid gametes. These gametes then combine to form a diploid plant. The diploid plant, the **sporophyte**, produces haploid spores by meiosis.

Take a look at the figure above. When the haploid gametes—eggs and sperms—unite during fertilization, a zygote is formed. This diploid organism, or sporophyte, begins the **sporophyte generation**. The zygote divides by mitosis and develops into a multicellular embryo that is supported and protected by the gametophyte plant.

The sporophyte plant divides and produces spores through meiosis. The spores, which are haploid, represent the first stage in the **gametophyte generation**. Each spore is capable of growing into a multicellular gametophyte plant. Once a gametophyte plant is produced the cycle continues again.

Here's an example of the life cycle of a moss:

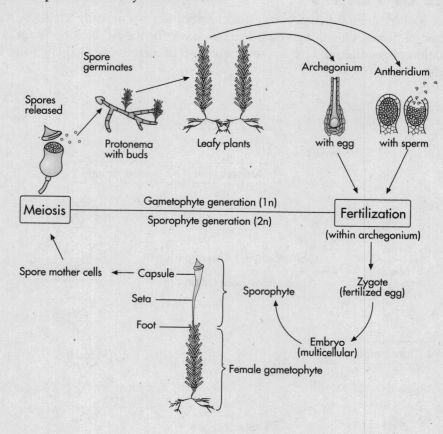

Notice from the bottom half of the diagram that the sporophyte plant grows on top of the gametophyte plant and depends on it for its survival.

Fortunately, you don't have to know all the details of this process. What you do have to know is that there is a major difference between the two phyla:

- In bryophytes, the *gametophyte* stage is dominant.

- In tracheophytes, the *sporophyte* stage is dominant and the gametophyte generation is reduced in size.

PLANT GROWTH

How do plants grow? Plants have unspecialized, actively dividing cells called **meristems**. The meristems of a plant make it grow in two ways. These two means of growth are called **primary growth** and **secondary growth**.

Primary growth increases the length of a plant. The tissues that cause primary growth are the **apical meristems** and are located in the tips of roots and stems.

Secondary growth is carried out by the **lateral meristems**. These dividing cells increase the girth, or width, of a plant, and are located on the sides of stems and roots. The lateral meristem produces two types of cells: **vascular cambium** and **cork cambium**. The vascular cambium produces secondary xylem and secondary phloem, which replace primary xylem and primary phloem. The cork cambium produces the tissues of the outer bark. Plants also have **lenticels**, which allow for gas exchange through the bark.

When it comes to plant growth, make sure you know the difference between the two types of growth and which tissues are responsible for each one.

SPECIAL STRUCTURES IN PLANTS

Roots

The growing root includes three regions: the **root tip**, the **elongation region,** and the **maturation region**. The root tip and elongation regions are the sites of ongoing **primary growth**. The **root apical meristem** includes tiny, undifferentiated cells that continually divide and form the zone of cell division. As cell division in the root apical meristem continues, the new cells left behind grow rapidly in length and push the root tip along. As the cells absorb water, elongation occurs. Tiny **root hairs**, extensions of the epidermal cells, form and provide an increased surface area through which water and dissolved minerals can move into the plant. In the maturation zone, vascular tissue forms primary xylem and phloem, which forms the **stele** (the inner concentric cylinder). Surrounding the primary xylem and phloem is a cylinder of cells called the **pericycle** (undifferentiated meristematic tissue), which gives rise to lateral roots.

Leaves

We already know that leaves play an important role in photosynthesis. But did you know that leaves are sometimes modified for other purposes?

Here's a list of some of the other functions of leaves:

1. Leaves can be modified to form **spines**, as in a cactus. This adaptation is great for protection.

2. Leaves can be adapted for water storage. Such fleshy leaves allow plants to survive particularly harsh environments where the water supply is intermittent or undependable.

3. Leaves can also be modified to trap prey. Insectivorous plants have specialized leaves that digest insects. Since they grow in soils deficient of essential nutrients, especially nitrogen, these plants are forced to eat insects. There are basically two general forms of these adaptations:

    -   Some leaves have tiny hairs that act like bear traps. For example, an insect brushing against the hairs in a **Venus flytrap** triggers the leaves to snap shut.

    -   Other leaves are adapted to form a "slippery slope," which traps insects. In a **Pitcher plant**, for example, once an insect gets inside, it can't get out. It slips down into the bell-shaped interior of the leaf, where it drowns in a mixture of water and enzymes. These enzymes then finish the job by digesting the insect.

FLOWERING PLANTS

When it comes to plant structures, one they're bound to test on the AP Biology Exam is the flower. You don't need to know everything about flowers for the test. Let's take a look at what you do need to know.

Flowering plants have several organs: the **stamen, pistil, sepals,** and **petals**.

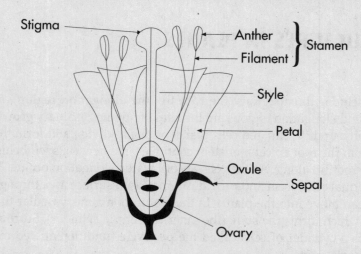

The male parts are collectively called the stamen, and the female parts are called the pistil. The sepals are the green, leaf-like structures that cover and protect the flower, while the brightly colored petals attract potential pollinators. Let's review each of these structures.

The stamen

The stamen consists of the **anther** and the **filament**. The anther is the structure that produces pollen grains. These **pollen grains**, called **microspores**, are the plant's male gametophytes or sperm cells. Pollen grains are produced and released into the air. The filament is the thin stalk that holds up the anther.

The pistil

The pistil includes three structures: the **stigma, style,** and **ovary**. The stigma is the "sticky" portion of the pistil that captures the pollen grains. The style is a tube-like structure that connects the stigma with the ovary.

The ovary is where fertilization occurs. Within the ovary are the **ovules,** which contain the plant's equivalent of the female gametophytes. In a fertilized plant the ovary develops into the fruit. Apples, pears, and oranges are all fertilized ovaries of flowering plants. The female gametes of plants are known as **megaspores**. They undergo meiosis to produce eight female nuclei, including one **egg nucleus** and two **polar nuclei**.

DOUBLE FERTILIZATION

Now that we've seen both the male and the female organs in plants, let's take a look at how they actually reproduce. Flowering plants carry out a process called **double fertilization**. When a pollen grain lands on the stigma, it germinates and grows a thin pollen tube down the style, which meets up with the ovary. The pollen grain then divides into two **sperm nuclei**. One sperm nucleus (n) fuses with an egg nucleus (n) to form a zygote (2n). This zygote will eventually form a plant. The other sperm nucleus (n) will fuse with two polar nuclei (2n) in the ovary to form the **endosperm** (3n). The endosperm will not develop into a plant. Rather, it will serve as food for the plant embryo. Double fertilization produces two things: a plant and food for the plant.

Let's review the steps involved in double fertilization:

- Grains of pollen fall onto the stigma. The pollen grains grow down the style into the ovary.

- The pollen grains (microspores) meet up with megaspores in the ovule. Microspores fertilize the megaspores.

- One microspore unites with an egg nuclei and eventually develops into a complete plant.

- The other microspore unites with two polar nuclei and develops into food for the plant, often in the form of a fruit.

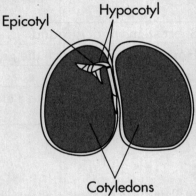

Early Seedling Development

As the embryo germinates, different parts of the plant begin to develop. The **cotyledons** are the first embryo leaves to appear. They temporarily store all the nutrients for the plant. The **epicotyl** is the part at the tip of the plant. This portion becomes the stems and leaves. The **hypocotyl** is the stem below the cotyledons. This portion becomes the roots of the plant. In some embryos, root development begins early, and the well-defined embryonic root is referred to as a **radicle**.

What triggers sexual reproduction in plants? Plants flower in response to changes in the amount of daylight and darkness. This is called **photoperiodism**. Plants fall into three main groups: **short-day plants, long-day plants,** and **day-neutral plants**. Although you'd think that plants bloom based on the amount of sunlight, they actually flower according to the amount of uninterrupted darkness.

Short-day plants require a long period of darkness, whereas long-day plants need short periods of darkness. Short-day plants usually bloom in late summer or fall when daylight is decreasing. Long-day plants, on the other hand, flower in late spring and summer when daylight is increasing. Day-neutral plants don't flower in response to daylight changes at all. They use other cues such as water or temperature.

The light receptor involved in photoperiodism is a pigment called **phytochrome**. In short-day plants, it *inhibits* flowering, whereas in long-day plants it *induces* flowering.

VEGETATIVE PROPAGATION

Flowering plants don't always reproduce via fertilization. In some cases, flowering plants can reproduce asexually. This process is known as **vegetative propagation**. That means we can take parts of the parent plant—such as the roots, stems, or leaves—to produce another plant. Some examples include **tubers, runners,** and **bulbs.** For instance, suppose you wanted to make white potatoes without fertilization. All you'd have to do is cut out the "eyes" of a potato, the tubers, and plant them. Each of the eyes will develop into a separate potato plant. **Grafting** is another way plants can be reproduced asexually.

Here's a list of the different types of vegetative propagation:

TYPES OF VEGETATIVE PROPAGATION		
Types of Vegetative Propagation	Description	Examples
Bulbs	Short stems underground	Onions
Runners	Horizontal stems above the ground	Strawberries
Tubers	Underground stems	Potatoes
Grafting	Cut a stem and attach it to a closely related plant	Seedless oranges

TROPICAL TROPISMS

Plants need light. Notice that all the plants in your house tip toward the windows. This movement toward the light is known as **phototropism**. As you also know, plants generally grow up and down: The branches grow upward, while the roots grow downward into the soil, seeking water. This tendency to grow toward or away from the earth is called **gravitropism**. All of these tropisms are examples of behavior in plants.

A **tropism** is a turning response to a stimulus.

There are three basic tropisms in plants. They're easy to remember because their prefixes indicate the stimuli to which plants react:

1. **Phototropism** refers to how plants respond to sunlight. For example, plants always bend toward light.

2. **Gravitropism** refers to how plants respond to gravity. Stems exhibit negative gravitropism (i.e., they grow up, away from the pull of gravity), whereas roots exhibit positive gravitropism (i.e., they grow downward into the earth).

3. **Thigmotropism** refers to how plants respond to touch. For example, ivy grows neatly about a post or trellis. This response is known as thigmotropism.

These responses are initiated by hormones. The major plant hormones you need to know belong to a class called **auxins**. Auxins serve many functions in plants. They can promote growth on one side of the plant. For example, in phototropism, the side of the plant that faces away from the sunlight grows faster, thanks to the plant's auxins, making the plant bend toward the light.

Generally speaking, auxins are in the tip of the plant, since this is where most growth occurs. Auxins are also involved in cell elongation and fruit development.

Other plant hormones that regulate the growth and development of plants are **gibberellins, cytokinins, ethylenes,** and **abscisic acid.** Here's a summary of the functions of plant hormones:

FUNCTIONS OF PLANT HORMONES	
Hormone	**Function**
Gibberellins	Promote stem elongation, especially in dwarf plants
Cytokinins	promote cell division and differention
Ethylene	Induces leaf abscission and promotes fruit ripening
Abscisic acid	Inhibits leaf abscission and promotes bud and seed dormancy
Auxins	Promote plant growth and phototropism

KEY WORDS

bryophytes
tracheophytes
vascular tissues
mosses
liverworts
xylem
phloem
tracheids
vessel elements
roots
root hairs
phloem vessels
sieve tube elements
companion cells
gymnosperms
angiosperms
perennial
tree rings
monocots
dicots
cotyledon
alternation of generation
gametophyte generation
sporophyte generation

meristems
primary growth
secondary growth
apical meristem
lateral meristem
vascular cambium
cork cambium
lenticels
spines
Venus flytrap
Pitcher plant
stamen
pistil
sepals
petals
anther
filament
microspores (pollen grains)
stigma
style
ovary
ovule
megaspores
egg nucleus

polar nuclei
double fertilization
sperm nuclei
endosperm
photoperiodism
short-day plants
long-day plants
day-neutral plants
phytochrome
vegetative propagation
tubers
runners
bulbs
grafting
phototropism
gravitropism
tropisms
thigmotropism
auxins
gibberellins
cytokinins
ethylene
abscisic acid

QUIZ

<u>Directions:</u> Each of the questions or incomplete statements below is followed by five suggested answers or completions. Select the one that is best in each case. Answers can be found on page 243.

1. All of the following are major characteristics of dicots EXCEPT:

   (A) They contain vascular tissue.
   (B) The leaves have netted veins.
   (C) They have two cotyledons.
   (D) The flower parts are in multiples of three.
   (E) The cotyledon provides food for the germinating embryo.

2. Conifers and flowering plants are classified in the same phylum. Which of the following characteristics do they share?

   (A) They are perennials.
   (B) The xylem is dead at maturity.
   (C) They are all woody plants.
   (D) They both contain seeds.
   (E) They both shed their leaves.

3. Which of the following structures is NOT part of the pistil?

   (A) Ovule
   (B) Ovary
   (C) Style
   (D) Anther
   (E) Stigma

4. The plant tissue that gives girth to a plant during each growing season is called the

   (A) phloem
   (B) tracheid
   (C) secondary xylem
   (D) lateral meristem
   (E) apical meristem

5. Carnivorous plants are most likely to be found in which of the following environments?

   (A) A nutrient-laden tropical rainforest canopy
   (B) An arid desert
   (C) A nitrogen-poor swamp
   (D) A temperate deciduous forest
   (E) A tropical savannah

6. All of the following are examples of tracheophytes EXCEPT:

  (A) trees
  (B) moss
  (C) grass
  (D) corn
  (E) beans

7. The plant hormone responsible for the ripening of fruit is

  (A) auxin
  (B) cytokinin
  (C) gibberellin
  (D) ethylene
  (E) abscisic acid

Directions: Each group of questions consists of five lettered headings followed by a list of numbered phrases or sentences. For each numbered phrase or sentence, select the one heading that is most closely related to it and fill in the corresponding oval on the answer sheet. Each heading may be used once, more than once, or not at all in each group.

Questions 8–12

  (A) Phytochrome
  (B) Phototropism
  (C) Photoperiodism
  (D) Thigmotropism
  (E) Gravitropism

8. Plant growth toward a light source

9. When a Venus flytrap snaps shut in response to an insect

10. The photoreceptor pigment responsible for sexual reproduction in plants

11. Growth of a plant in response to the direction of gravity

12. Biological effects caused by changes in day length

11

Animal Structure and Function

To carry on with the business of life, higher organisms must all contend with the same basic challenges: obtaining nutrients, distributing them throughout their bodies, voiding wastes, responding to their environments, and reproducing. To accomplish these basic tasks, nature has come up with solutions. However, for all their differences, most animals have remarkably similar ways of dealing with these challenges.

This chapter looks at the basic structures of animals and the ways in which they function. Since the AP Biology Exam includes many questions on human anatomy and physiology, we'll focus primarily on how these systems have evolved in human beings.

The systems we'll look at include:

- The digestive system
- The respiratory system
- The circulatory system

- The immune system
- The excretory system
- The nervous system
- The musculoskeletal system
- The endocrine system
- The reproductive system
- Morphogenesis, or "development"

I: THE DIGESTIVE SYSTEM

All organisms need nutrients to survive. But where do the nutrients—the raw building blocks—come from? That depends on whether the organism is an autotroph or heterotroph. As you may recall, autotrophs make their own food through photosynthesis, and all of the building blocks come from their immediate environment: CO_2, water, and sunlight. Heterotrophs, on the other hand, can't make their own food. They must acquire their energy from outside sources.

When we talk about digestion, we're talking about the breakdown of large food molecules into simpler compounds. These molecules are then absorbed by the body to carry out cell activities. In fact, everything we'll discuss in this section revolves around three simple questions:

1. What do organisms need from the outside world in order to survive?
2. How do they get those things?
3. What do they do with them once they get them?

Multicellular organisms have come up with a variety of ways of getting their nutrients. In simple animals, food is digested through **intracellular digestion**—that is, digestion occurs within food vacuoles. For example, a hydra encloses the food it captures in a food vacuole. Lysosomes containing digestive enzymes then fuse with the vacuole and break down the food. More complex animals have evolved a digestive tract and digest food through **extracellular digestion**. That is, the food is digested in a gastrovascular cavity. For example, in grasshoppers, food passes through specialized regions of the gut: the **mouth**, **esophagus**, **crop** (a storage organ), **stomach**, **intestine**, **rectum** and **anus**.

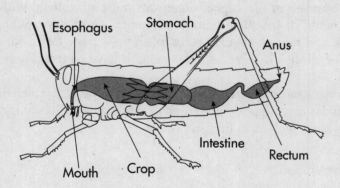

THE HUMAN DIGESTIVE SYSTEM

The human digestive tract consists of the **mouth, esophagus, stomach, small intestine, large intestine**, and **accessory organs** (liver, pancreas, gall bladder, and salivary glands). Four groups of molecules must be broken down by the digestive tract: **starch, proteins, fats**, and **nucleic acids.**

The mouth

The first stop in the digestive process is the mouth, or **oral cavity**.

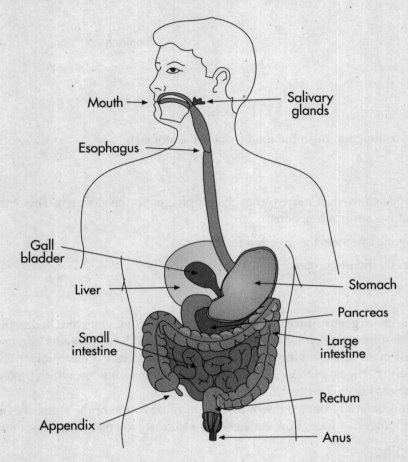

When food enters the mouth mechanical and chemical digestion begins. The chewing, softening and breaking up food is called mechanical digestion. The mouth also has **saliva** in it. Saliva, which is secreted by the **salivary glands**, contains an important enzyme known as **salivary amylase**. Salivary amylase begins the chemical digestion of *starch* into maltose.

Once chewed, the food, now a ball called a **bolus**, moves through the **pharynx** and into the **esophagus**. Food moves through the esophagus in a wavelike motion known as **peristalsis**.

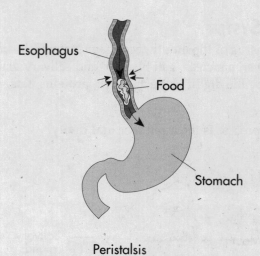

Esophagus

Food

Stomach

Peristalsis

The waves of contraction push the food toward the stomach.

The stomach

Once food has been chewed, it moves from the esophagus to the stomach. The stomach is a thick, muscular sac that has several functions:

- It temporarily stores the ingested food.
- It partially digests proteins.
- It kills bacteria.

The stomach secretes **gastric juices,** which contain digestive enzymes and hydrochloric acid (HCl). One of the most important enzymes is **pepsin**, which breaks down proteins into smaller peptides. Pepsin works best in an acidic environment. When HCl is secreted, it lowers the pH of the stomach and activates pepsin to digest proteins. The stomach also secretes mucus, which protects the stomach lining from the acidic juices. Finally, HCl kills bacteria.

Food is also mechanically broken down by the churning action of the stomach. Once that's complete, this partially digested food, now called **chyme**, is ready to enter the small intestine.

The small intestine

The small intenstine has three regions: the duodenum, the jejunum, and the ileum. The chyme moves into the first part of *small intestine*, the duodenum, through the **pyloric sphincter**. The small intestine is very long—about 23 feet in an average man. This is where all three food groups are completely digested. The walls of the small intestine secrete enzymes that break down proteins (peptidases) and carbohydrates (maltase, lactase, and sucrase).

The pancreas

The pancreas secretes a number of enzymes into the small intestine: **trypsin, chymotrypsin, pancreatic lipase**, and **pancreatic amylase**. Trypsin and chymotrypsin break down proteins into dipeptides. Pancreatic lipase breaks down lipids into fatty acids and glycerol. Pancreatic amylase, on the other hand, breaks down starch into disaccherides. Ribonuclease and deoxyribonyclease break down nucleic acids into nucleotides.

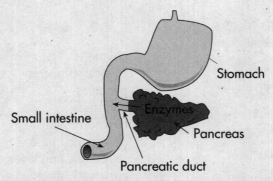

These enzymes are secreted into the small intestine via the **pancreatic duct**.

Another substance that works in the small intestine is called **bile**. Bile is not a digestive enzyme. It's an **emulsifier**, meaning that it mechanically breaks up fats into smaller fat droplets. This process makes the fat globules more accessible to pancreatic lipase. Bile enters the small intestine by the bile duct, which merges with the pancreatic duct.

Here's something you should memorize:

> Bile is made in the liver and stored in the gall bladder.

Once food is broken down, it is absorbed by tiny, fingerlike projections of the intestine called **villi** and **microvilli**. Villi and microvilli are folds that increase the surface area of the small intestine for food absorption. Within each of the villi is a capillary that absorbs the digested food and carries it into the bloodstream. Within each villus there are also lymph vessels, called **lacteals**, which absorb fatty acids.

Don't forget that hormones are also involved in the digestive system: **gastrin** (which stimulates stomach cells to produce gastric juice), **secretin** (which stimulates the pancreas to produce bicarbonate and digestive enzymes), and **cholecystokinin** (which stimulates the secretion of pancreatic enzymes and the release of bile).

Here's a summary of the pancreatic enzymes:

THE PANCREATIC ENZYMES AND THE FOODS THEY DIGEST	
Pancreatic Enzymes	**Food Substance**
pancreatic amylase	starch
pancreatic lipase	fat
trypsin, chymotrypsin	protein
proteolytic enzymes maltase, lactase	proteins carbohydrates

The large intestine

The large intestine is much shorter and thicker than the small intestine. The large intestine has an easy job: It reabsorbs water and salts. The large intestine also harbors harmless bacteria that are actually quite useful. These bacteria break down undigested food and in the process provide us with certain essential vitamins, like **Vitamin K**. The leftover undigested food, called **feces**, then moves out of the large intestine and into the **rectum**.

KEY WORDS

intracellular digestion
extracellular digestion
mouth
esophagus
crop
small intestine
large intestine
anus
accessory organs
starch
proteins
fats
oral cavity
mastication
saliva
salivary glands
salivary amylase
bolus
pharynx
peristalsis
gastric juices
pepsin

chyme
pyloric sphincter
pancreas
trypsin
chymotrypsin
pancreatic lipase
pancreatic amylase
pancreatic duct
bile
emulsifier
villi
microvilli
lacteals
liver
gall bladder
stomach
cholecystokinin
secretin
gastrin
Vitamin K
feces
rectum

II: THE RESPIRATORY SYSTEM

All cells need oxygen for aerobic respiration. For simple organisms, such as Platyhelminthes, no special structures are needed because the gases can easily diffuse across every cell membrane. In other multicellular organisms, however, the cells are not in direct contact with the environment. These organisms must find other ways of getting oxygen into their systems. For some animals, such as segmented worms, gas exchange occurs directly through their skin. Others, such as insects, have special tubes called **tracheae**. Air enters these tubes through tiny openings called **spiracles**. Among vertebrates, the respiratory structures you should be familiar with are **gills** (used by many aquatic creatures) and **lungs**.

THE HUMAN RESPIRATORY SYSTEM

Let's talk about how air gets into the body. Air enters through the nose or mouth:

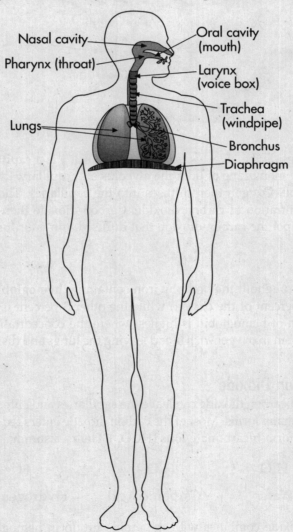

The nose cleans, warms, and moistens the incoming air and passes it through the **pharynx** and **larynx** (voicebox). Next, air enters the **trachea**. A special flap called the **epiglottis** covers the trachea when you swallow, preventing food from going down the wrong pipe. The trachea also has cartilage rings to help keep the air passage open as air rushes in.

The trachea then branches into two bronchi: the **left bronchus** and the **right bronchus**. These two tubes service the lungs. In the lungs, the passageways break down into smaller tubes known as **bronchioles**. Each bronchiole ends in a tiny air sac known as an **alveolus**. These sacs enable the lungs to have an enormous surface area: about 100 square meters. Let's take a look at one of these tiny air sacs:

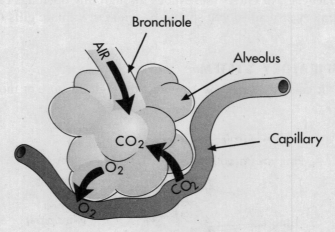

You'll notice that alongside the alveolus (singular of "alveoli") is a **capillary**. Oxygen and carbon dioxide diffuse across the membrane of both the alveolus and capillary. Every time you inhale, you send oxygen to the alveoli. Oxygen then diffuses into the capillaries. The capillaries, on the other hand, have a high concentration of carbon dioxide. Carbon dioxide then diffuses into the alveoli. When you exhale, you expel the carbon dioxide that diffused into your lungs. Gas exchange occurs via passive diffusion.

Transport of Oxygen

Oxygen is transported throughout the body by iron-containing **hemoglobin** in red blood cells. Hemoglobin transports 97 percent of the oxygen while the other 3 percent is dissolved in the plasma. The percent O_2 saturation of hemoglobin is highest where the concentration of oxygen is greatest. Oxygen binds to hemoglobin in oxygen-rich blood leaving the lungs and dissociates from hemoglobin in oxygen-poor tissues

The Transport of Carbon Dioxide

We've just mentioned that carbon dioxide can leave the capillaries and enter the lungs. However, carbon dioxide can travel in many forms. Most of the carbon dioxide enters red blood cells and combines with water to eventually form **bicarbonate ions** (HCO_3^-). Here's a summary of the reaction:

$$CO_2 \quad + \quad H_2O \quad \leftrightharpoons \quad H_2CO_3 \quad \rightleftharpoons \quad H^+ \quad + \quad HCO_3^-$$

Carbon dioxide   Water   Carbonic Acid   Hydrogen ion   Bicarbonate ion

Sometimes carbon dioxide combines with the amino group in hemoglobin, an iron-containing protein in red blood cells and is transported this way to the lungs or it mixes with plasma.

The Mechanics of Breathing

What happens to your body when you take a deep breath? Your diaphragm and intercostal muscles contract and your ribcage expands. This action increases the volume of the lungs, allowing air to rush in. This process of taking in oxygen is called **inspiration**. When you breathe out and let carbon dioxide out of your lungs, that's called **expiration**. Your respiratory rate is controlled by **chemoreceptors**. As your blood pH decreases, chemoreceptors send nerve impulses to the diaphragm and intercostal muscles to increase your respiratory rate.

KEY WORDS

trachea	bronchioles
spiracles	alveolus
gills	capillary
lungs	hemoglobin
pharynx	bicarbonate ions
larynx	inspiration
epiglottis	expiration
left bronchus	chemoreceptors
right bronchus	diaphragm

III: THE CIRCULATORY SYSTEM

Most organisms need to carry out two tasks: (1) supply their bodies with nutrients and oxygen, and (2) get rid of wastes. Many simple aquatic organisms have no trouble moving materials across their membranes since their metabolic needs are met by diffusion. Larger organisms, on the other hand, particularly terrestrial organisms, can't depend on diffusion. They therefore need special circulatory systems to accomplish internal transport.

There are two types of circulatory systems: an **open circulatory system** and a **closed circulatory system**. In an open circulatory system, blood is carried by open-ended blood vessels that spill blood into the body cavity. In arthropods, for example, blood vessels from the heart open into internal cavities known as **sinuses**. Other organisms have closed circulatory systems. That is, blood flows continuously through a network of blood vessels. Earthworms and some mollusks have a closed circulatory system as do vertebrates.

THE HUMAN CIRCULATORY SYSTEM

The heart is divided into four chambers, two on the left and two on the right. The four chambers of the heart are the **right atrium**, the **right ventricle**, the **left atrium**, and the **left ventricle**. Let's take a look at a picture of the heart:

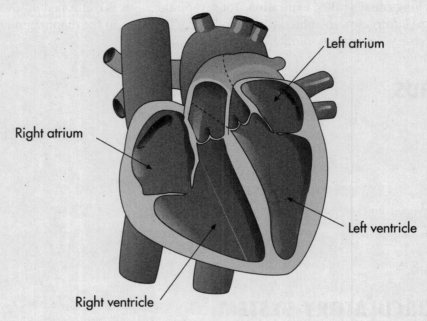

Left atrium

Right atrium

Left ventricle

Right ventricle

The heart pumps blood in a continuous circuit. Since blood makes a circuit in the body, it doesn't matter where we begin to trace the flow of blood. For our purposes, we'll begin at the point in the circulatory system where the blood leaves the heart and enters the body: the left ventricle. When blood leaves the left ventricle it will make a tour of the body. We call this systemic circulation.

Systemic circulation

Blood leaves the heart through the **aortic semilunar valve** and enters a large blood vessel called the **aorta**. The aorta is the largest artery in the body. The aorta then branches out into smaller vessels called **arteries**.

Arteries always carry blood away from the heart. Just remember "A" stands for "away" from the heart. They're able to carry the blood because arteries are thick-walled, elastic vessels. The arteries become even smaller vessels called **arterioles** and then the smallest vessels called **capillaries**.

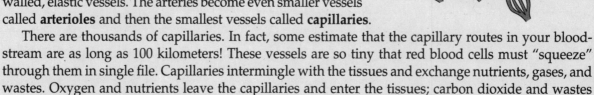

Artery

Artery Arteriole

Capillaries

There are thousands of capillaries. In fact, some estimate that the capillary routes in your bloodstream are as long as 100 kilometers! These vessels are so tiny that red blood cells must "squeeze" through them in single file. Capillaries intermingle with the tissues and exchange nutrients, gases, and wastes. Oxygen and nutrients leave the capillaries and enter the tissues; carbon dioxide and wastes leave the tissues and enter the capillaries.

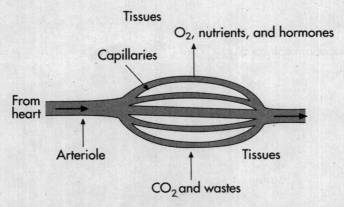

Before we take a look at the next stage of circulation, let's recap the pathway of blood through the body:

- Blood leaves the heart's left ventricle via the aorta.

- It travels through the arteries to the arterioles, and eventually to the capillaries.

- Gas and nutrient/waste exchange occurs between the blood and the cells through the capillary walls.

Back to the heart

After exchanging gases and nutrients with the cells, blood has very little oxygen left. Most of its oxygen was donated to the cells through the capillary walls. Since the blood is now depleted of oxygen, it is said to be **deoxygenated**. To get a fresh supply of oxygen the blood now needs to go to the lungs.

But the blood doesn't go *directly* to the lungs. It must first go back to the heart. As the blood returns to the heart, the vessels get bigger and bigger

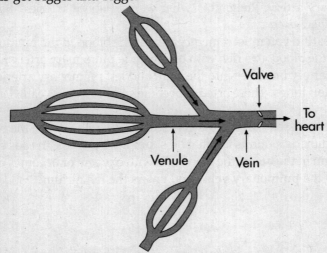

From the capillaries, blood travels through vessels called **venules** and then through larger vessels called **veins**. Veins always carry blood *toward* the heart. Veins are thin-walled vessels with valves that prevent the backward flow of blood.

Blood eventually enters the heart's right atrium via two veins known as the **superior vena cava** and the **inferior vena cava.**

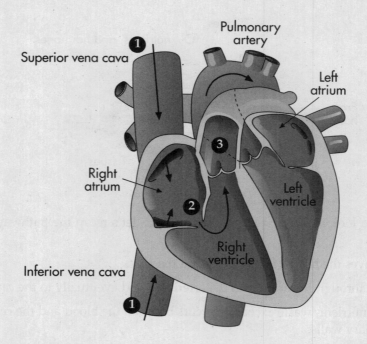

Superior vena cava

Pulmonary artery

Left atrium

Right atrium

Left ventricle

Inferior vena cava

Right ventricle

Blood now moves through the heart. Blood travels from the right atrium to the right ventricle through the **right atrioventricular valve** (or **tricuspid**). From the right ventricle, blood will go out again into the body, but this time towards the lungs. This is called the **pulmonary circulation**.

The pulmonary system

Blood leaves the right ventricle through the pulmonary semilunar valve and enters a large artery known as the **pulmonary artery**. Remember what we said about arteries? Blood vessels that leave the heart are always called *arteries*.

There's one major feature you must remember about the blood in the pulmonary system. Whereas in systemic circulation the blood was rich with oxygen, the pulmonary artery is carrying *deoxygenated* blood. The pulmonary artery branches into the right and left pulmonary arteries, which lead, respectively, to the right and left lungs. These arteries become smaller vessels called *arterioles* and then once again *capillaries*.

We just said that these vessels carry deoxygenated blood. In the lungs, the blood will pick up oxygen and dump carbon dioxide. Sounds familiar? It should. It's just like the gas exchange we discussed in the respiratory system. In the lungs, the blood fills with oxygen, or becomes **oxygenated**. The blood returns to the heart via the **pulmonary veins** and enters the left atrium.

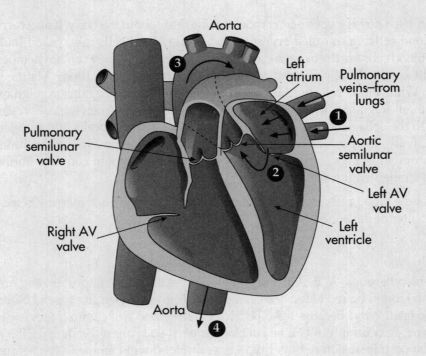

Blood then moves to the left ventricle through the left atrioventricular valve (or bicuspid or mitral valve). Now we've completed our tour of the heart. Let's recap the events in pulmonary circulation:

- Deoxygenated blood leaves the right ventricle via the pulmonary artery.

- The pulmonary artery branches into the right and left pulmonary arteries, carrying the blood to the lungs.

- Blood travels from the arteries to the arterioles, and eventually to the capillaries.

- Gas exchange occurs between the capillaries and alveoli in the lungs.

- Once the blood is oxygenated, it returns to the heart through the pulmonary veins.

HEART CYCLE

Your heart contracts and relaxes automatically, about 72 times a minute. To make sure that your heart beats rhythmically, there is a special conduction system. The beat begins in tissues in the right atrium called the **sinoatrial (SA) node** ("the pacemaker"). The impulse then spreads through both atria and conducts directly to the **atrioventricular (AV) node**. From the AV node, the action potential spreads to the **bundle of His** and then to the **Purkinje fibers** in the walls of both ventricles. This generates a strong contraction. The part of the cycle in which contraction occurs is called **systole** and the part in which relaxation occurs is called **diastole**.

THE CONTENTS OF BLOOD

Now let's take a look at blood itself. Blood consists of two things:

- Fluid (called **plasma**)

- Cells and cell fragments suspended in the fluid

Blood doesn't just carry oxygen and carbon dioxide throughout the body. It also carries three types of cells: **red blood cells** (also called **erythrocytes**), **white blood cells** (also called **leukocytes**), and **platelets**. Red blood cells are the oxygen-carrying cells in the body. They contain hemoglobin, the protein that actually carries the oxygen (and carbon dioxide) throughout the body. Mature red blood cells lack a nucleus. White blood cells fight infection by fighting the body against foreign organisms.

Platelets are cell fragments that are involved in blood clotting. When a blood vessel is damaged, platelets stick to the collagen fibers of the vessel wall. The damaged cells and platelets release substances that activate clotting factors and a series of reactions occur. First, a prothrombin activator converts prothrombin (a plasma protein) to thrombin. Then thrombin converts fibrinogen to fibrin threads, which strengthen the clot.

All of the blood cells are made in the bone marrow. The bone marrow is located in the center of the bones.

Blood types

There are four blood groups: **A**, **B**, **AB**, and **O**. Blood types are pretty important and are based on the type of antigen(s) found on red blood cells. If a patient is given the wrong type of blood in a transfusion, it could be fatal! Why? Because red blood cells in the blood will clump if they are exposed to the wrong blood type. For example, if you've got blood type A, i.e., your red cells carry the A antigen, and you receive a blood transfusion of blood type B, your blood will clump. That's because your blood contains **antibodies**, an immune substance that will bind and destroy the foreign blood.

What is important to remember about the different blood types is that type O blood is the universal donor and that type AB is the universal recipient. This means that anyone can receive a blood transfusion of type O blood, while those with type AB blood (which is very rare among Americans—only about 4% of the population) can receive any kind of blood without risk. Rh factors are also antigens found on red blood cells. Persons with these antigens are Rh^+ and those without them are Rh^-.

The Lymphatic System

In addition to the circulatory system, vertebrates have another system called the lymphatic system. The **lymphatic system** is made up of a network of vessels that conduct **lymph**, a clear, watery fluid. Lymph vessels are found throughout the body along the routes of blood vessels (see the illustration on the next page).

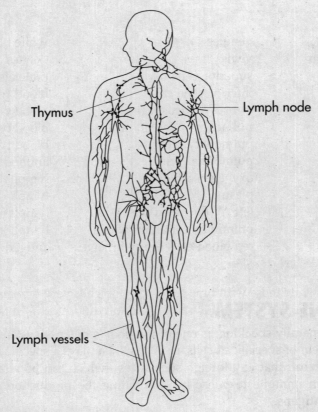

Thymus

Lymph node

Lymph vessels

The lymphatic system has two functions:

- It collects, filters and returns fluid to the blood by the contraction of adjacent muscles.

- It fights infection using lymphocytes, cells found in lymphnodes.

Sometimes a lymph vessel will form a **lymph node**, a bulge found along the course of a lymph vessel. A lymph node contains lots of lymphocytes:

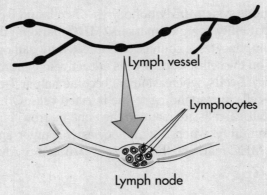

Lymph vessel

Lymphocytes

Lymph node

Lymphocytes are important in fighting infection. They multiply rapidly when they come in contact with an **antigen**, or foreign protein. The lymph nodes swell when they're fighting an infection. That's why when you have a sore throat one of the first things a doctor does is touch the sides of your throat to see if your lymph nodes are swollen, a probable sign of infection.

KEY WORDS

open circulatory system
closed circulatory system
sinuses
right atrium
right ventricle
left atrium
left ventricle
systemic circulation
aortic semilunar valve
aorta
arteries
arterioles
capillaries
deoxygenated blood

venules
veins
superior vena cava
inferior vena cava
right AV (tricuspid)valve
pulmonary circulation
pulmonary artery
pulmonary vein
oxygenated blood
plasma
left AV (bicuspid) valve
pulminary semilunar valve
red blood cells or erythrocytes

white blood cells or leukocytes
platelets
blood type A
blood type B
blood type AB
blood type O
antibodies
lymphatic system
lymph
lymph node
lymphocytes
antigen

IV: THE IMMUNE SYSTEM

The immune system, generally speaking, is one of the body's defense systems. It is a carefully and closely coordinated system of specialized cells, each of which plays a specific role in the war against bodily invaders. You'll recall that any foreign substances—which can be viral, bacterial, or simply chemical—can trigger an immune response by stimulating the production of antibodies that are known collectively as **antigens**.

The body's first line of defense is the skin and mucous lining of the respiratory and digestive tracts. Once the pathogens break through this line of defense, other nonspecific defense mechanisms are activated. These include **phagocytes** (which engulf pathogens), **complement proteins** (which lyse the cell wall of the pathogen), **interferons** (which inhibit viral replication and activate surrounding cells that have antiviral actions), and **inflammatory response** (a series of events in response to pathogen invasion or physical injury).

TYPES OF IMMUNE CELLS

The primary cells of the immune system are lymphocytes—T-cells and B-cells. The plasma membrane of cells has major histocompatibility complex markers (**MHC markers**) that distinguish between self and nonself cells. When T-lymphocytes encounter cells infected by pathogens, they recognize the foreign antigen-MHC markers on the cell surface. T-cells are activated, multiply, and give rise to clones. Some T-cells become **memory T-cells**, whereas others become **helper T-cells**. Helper T-cells activate B-lymphocytes and other T-cells in responding to the infected cells. Other T-cells, **cytotoxic T-cells**, recognize and kill infected cells. T-cells are made in the bone marrow, but mature in the thymus.

In antibody-mediated immunity, when B-lymphocytes encounter antigen-presenting cells (like **macrophages**) with foreign MHC markers, they are activated and produce clones. Some B-cells become memory cells. Other B-cells become plasma cells and produce antibodies, which combine with antigens. Helper T-cells are also involved and produce interleukins. Both memory T- and memory B-cells are responsible for long-term immunity.

AIDS

AIDS, or "acquired immunodeficiency syndrome," is a devastating disease thst interferes with the body's immune system. AIDS essentially wipes out the white blood cells, preventing the body from defending itself. Those afflicted with AIDS do not die of AIDS itself but rather of infections that they can no longer fight off due to their ruined immune systems.

One thing to remember about the immune cells and blood cells: All blood cells, white and red, are produced in the **bone marrow**. To summarize:

- *T-lymphocytes* actually fight infection and help the B-lymphocytes proliferate.

- *B-lymphocytes* produce *antibodies*.

Now that you know everything you need to know about immunity for the AP test, let's talk about the excretory system.

KEY WORDS

immune system
antigens
MHC markers
B-lymphocytes
T-lymphocytes
cytotoxic T-cells
helper T-cells
macrophages
antibodies
AIDS
bone marrow

V: THE EXCRETORY SYSTEM

As you already know, all organisms must get rid of wastes. In this chapter, we'll focus primarily on how organisms get rid of **nitrogenous wastes**—products containing nitrogen and regulate water. When cells break down proteins, one of the by-products is **ammonia** (NH_3), a substance that is toxic to the body. Consequently, organisms had to develop ways of converting ammonia to a less poisonous substance. Some animals convert ammonia to **uric acid**, while others convert ammonia to **urea**. Some examples of excretory organs among invertebrates are **nephridia** (found in earthworms) and **Malpighian tubules** (found in arthropods).

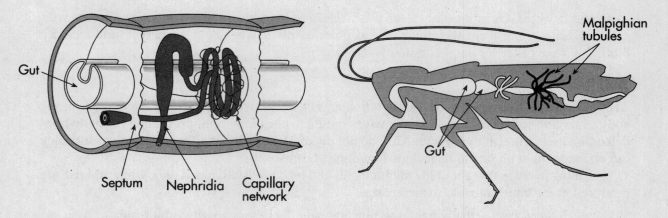

Gut

Malpighian tubules

Septum   Nephridia   Capillary network

Gut

THE HUMAN EXCRETORY SYSTEM

In humans, the major organ that regulates excretion is the **kidney**. Each kidney is made up of a million tiny structures called **nephrons**. Nephrons are the functional units of the kidney. A nephron consists of several regions: the **Bowman's capsule**, the **proximal convoluted tubule**, the **loop of Henle**, the **distal convoluted tubule**, and the **collecting duct**.

Here's what a nephron looks like under a microscope:

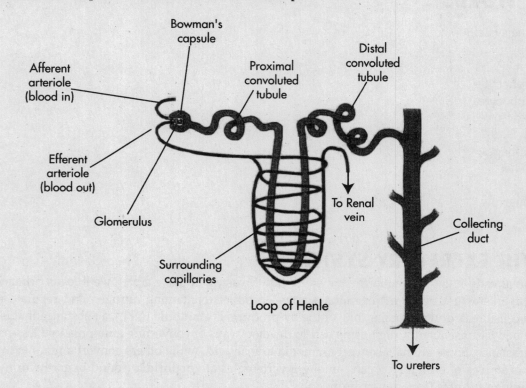

Bowman's capsule

Afferent arteriole (blood in)

Proximal convoluted tubule

Distal convoluted tubule

Efferent arteriole (blood out)

Glomerulus

To Renal vein

Collecting duct

Surrounding capillaries

Loop of Henle

To ureters

How does a nephron work? Let's trace the flow of blood in a nephron. Blood enters the nephron at the Bowman's capsule. A blood vessel called the **renal artery** leads to the kidney and branches into arterioles, then tiny capillaries. A ball of capillaries that "sits" within a Bowman's capsule is called a **glomerulus**. Blood is filtered as it passes through the glomerulus and the plasma is forced out of the capillaries into the the Bowman's capsule. This plasma is now called a **filtrate**.

The filtrate travels along the entire nephron. From the Bowman's capsule, the filtrate passes through the proximal convoluted tubule, then the loop of Henle, then the distal convoluted tubule, and finally the collecting duct. As it travels along the tube, the filtrate is modified to form urine.

What happens next? The concentrated **urine** moves from the collecting ducts into the **ureters**, then into the **bladder**, and finally out through the **urethra**.

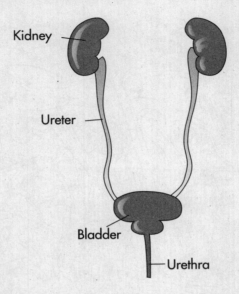

Kidney

Ureter

Bladder

Urethra

How urine is made

Urine is produced in the nephron by three processes: **filtration**, **reabsorption**, and **secretion**.

- **Filtration**—The blood is filtered as it passes through the glomerulus to the Bowman's capsule. Small substances, such as ions, water, glucose, and amino acids, easily pass through the capillary walls. Large substances, such as proteins and blood cells, are too large to pass through.

- **Reabsorption**—As the filtrate moves through the proximal convoluted tubule, some materials are reabsorbed. The small solutes, such as water, nutrients, and salts, leave the proximal convoluted tubule and are reabsorbed by a network of capillaries, the **peritubular capillaries** that surround the tubules. The material remaining in the tubule is urine.

- **Secretion**—As the filtrate moves through the convoluted tubules, some substances, such as H^+, potassium, and ammonium ions, are secreted from the surrounding capillaries into the tubule.

Hormones of the kidney

Two hormones regulate the concentration of water and salt in the kidneys: **vasopressin** (also known as **antidiuretic hormone**) and **aldosterone**. Antidiuretic hormone (ADH) allows water to be reabsorbed from the collecting duct. If your fluid intake is low (or if you're dehydrated), ADH helps your body retain water by concentrating the urine. If, however, your fluid intake is high, ADH levels will be low, and the body won't reabsorb most of the water. Your urine will contain lots of water and therefore be dilute. For now, just remember that ADH controls the volume of urine. The other hormone, aldosterone, is responsible for regulating sodium reabsorption at the distal convoluted tubule.

Skin

The **skin** is also an excretory organ that gets rid of excess water and salts from the body. Believe it or not, your skin is the largest organ in the body! It contains 2.5 million sweat glands that secrete water and ions through pores. The skin's primary function is to regulate body temperature.

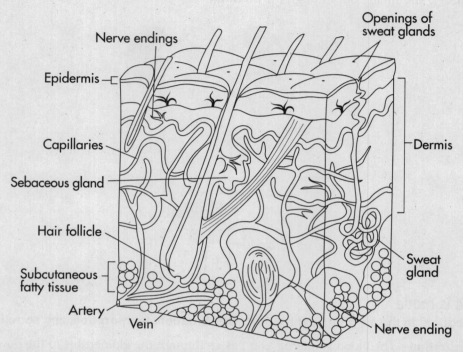

The skin has three layers: the **epidermis**, **dermis**, and **subcutaneous tissue** (or hyperdermis). Sweat glands are found in the dermis layer along with blood vessels, nerves, and sebaceous—or oil—glands.

The epidermis is covered by a layer of dead cells called the **stratum corneum**. These cells form a barrier against invading microorganisms. The bottom layer, the subcutaneous tissue, is mostly fats.

To summarize, in humans, two organs control fluid balance and dispose of metabolic wastes:

- The kidney—gets rid of nitrogenous wastes and reabsorbs water and salt.

- The skin—gets rid of excess salt and water.

KEY WORDS

nitrogenous wastes
ammonia
uric acid
urea
nephridia
Malpighian tubules
kidney
nephrons
Bowman's capsule
proximal convoluted tubule
loop of Henle
distal convoluted tubule
collecting duct
renal arteries
glomerulus

filtrate
ureters
bladder
urethra
filtration
reabsorption
secretion
vasopressin (antidiuretic hormone)
aldosterone
skin
epidermis
dermis
subcutaneous tissue
stratum corneum

VI: THE NERVOUS SYSTEM

All organisms must be able to react to changes in their environment. As a result, organisms have evolved systems that pick up and process information from the outside world. The task of coordinating this information falls to the nervous system. The simplest nervous system is found in the hydra. It has a **nerve net** made up of a network of nerve cells whose impulse travels in both directions. As animals became more complex, they developed clumps of nerve cells called **ganglia**. These cells are like primitive brains. The more complex organisms have a brain with specialized cells called **neurons**.

NEURONS

The basic unit of structure and function in the nervous system is a **neuron**. That's because neurons receive and send the neural impulses that trigger organisms' responses to their environments. Let's talk about the parts of a neuron. A neuron consists of a **cell body**, **dendrites**, and an **axon**.

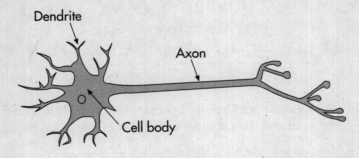

The cell body contains the nucleus and all the usual organelles found in the cytoplasm. Dendrites are short extensions of the cell body that receive stimuli. The axon is a long, slender extension that transmits an impulse from the cell body to another neuron or to an organ. A nerve impulse begins at the top of the dendrites, passes through the dendrites to the cell body, and moves down the axon.

Types of neurons

Neurons can be classified into three groups: **sensory neurons**, **motor (effector) neurons**, and **interneurons**. Sensory neurons receive impulses from the environment and bring them to the body. For example, sensory neurons in your hand are stimulated by touch. A motor neuron transmits the impulse to muscles or glands to produce a response. The muscle will respond by contracting or the gland will respond by secreting a substance (e.g., a hormone). Interneurons are the links between sensory neurons and motor neurons. They're found in the brain or spinal cord:

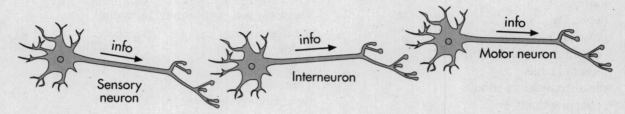

How neurons communicate

Before we talk about the events related to the transmission of a nerve impulse, let's review how neurons interact. There are billions of neurons running throughout the body, firing all the time. More often than not, one or more neurons are somewhat "connected." This means that one neuron has its dendrites next to another neuron's axon. In this way, the dendrites of one cell can pick up the impulse sent from the axon of another cell. The second neuron can then send the impulse to its cell body and down its axon, passing it on to yet another cell.

Resting potential

Neurons are not always firing away. Sometimes they're at rest. This simply means that the neuron is not sending or picking up any impulses. The transmission of an impulse depends on concentration gradients along the axon membrane. In a resting neuron, there is a high concentration of sodium ions (Na^+) outside the axon membrane and lots of potassium ions (K^+) inside the axon membrane:

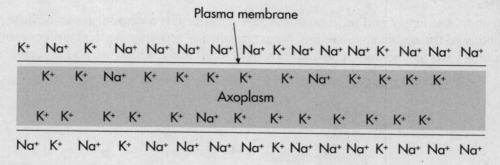

In fact, the outside has a lot more positive charges than the inside. This creates an overall negative charge on the inside of the axon and the neuronal membrane is said to be **polarized**.

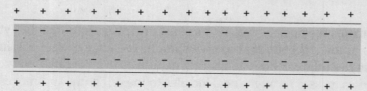

This negative charge within the axon is –70 millivolts (mV).

Action potential

Here's what a neuron does in response to a stimulus. If a stimulus has enough intensity to excite a neuron, the cell reaches its **threshold**—the minimum amount of stimulus a neuron needs to respond. This creates what we call an **action potential**, that is, a change in the membrane potential that produces a nerve impulse. The action potential is an **all-or-none response**—it doesn't fire "part way."

Depolarization

At the point where the axon connects to the cell body, tiny gated sodium ion channels open up and allow sodium ions to rush into the cell. So many sodium ions rush in that the cell now becomes more positive inside than outside. This change is known as **depolarization**: The interior of the cell has "switched" its polarity from a negative to a positive charge. For now, just remember that an action potential makes the cell depolarize.

The net change is substantial: The charge has shifted from the –70 mV we saw earlier to about +35 mV:

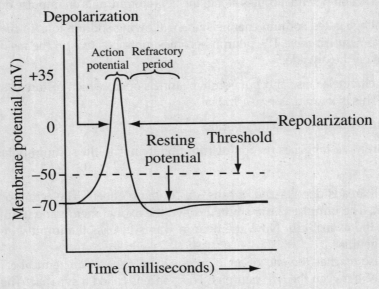

Let's recap what's happened so far. In an action potential:

- The cell's tiny "gates" open up.

- Sodium ions rush in.

- The "polarity" of the cell changes: The axon is now positive on the inside and negative on the outside.

Repolarization

Once sodium ions have "flooded" the neuron, the sodium channels close. At this point, the potassium channels open. The potassium ions, which are on the inside of the axon, now rush out. As the potassium ions move out of the cell, the electrical charges reverse again. The inside of the cell becomes more negative than the outside of the cell.

We can now say that this section of the neuron has been **repolarized**. In other words, the charge has returned to its original polarization.

The refractory period

Here's one thing you should remember: Although the charge has returned to its original state, at the end of the action potential the ions are now on the wrong side of the axonal membrane. Sodium ions are on the inside and potassium ions are on the outside of the axonal membrane. Originally, sodium ions were on the outside and potassium ions were on the inside. The neuron reestablishes the order of the ions and this process is carried out by the **sodium-potassium pump**.

This pump reestablishes the original ion distribution by kicking three sodium ions out of the cell for every two potassium ions it brings into the cell. The period after an action potential is known as the **refractory period**.

The sodium channels are now reset and are able to open but the cell membrane potential is farther from the threshold. A greater stimulus is required to reach the threshold so it is more difficult to initiate another action potential.

To summarize:

- A "resting" neuron is polarized, that is, it is more negative on the inside than on the outside.

- When an action potential comes along, the neuron transmits an impulse down its axon.

- First, voltage gated sodium channels open, allowing sodium ions to rush in. This is known as *depolarization*: The neuron becomes more positive on the inside and more negative on the outside.

- Sodium channels close and potassium channels open, which restores its negative charge. This is known as *repolarization*.

- The neuron enters a refractory period.

- The neuron reestablishes the ion distribution thanks to the sodium-potassium pump.

When one small area is depolarized, it causes a "domino effect." The action potential spreads to the rest of the axon. The impulse is transmitted down the axonal membrane until it reaches the end of the axon called the **axon bulb**. Now, the neuron wants to pass the impulse to the next neuron. How does it manage this?

When an impulse reaches the end of an axon, the axon releases a chemical called a **neurotransmitter** into the space between the two neurons. This space is called a **synapse**. The neurotransmitter diffuses across the synaptic cleft and binds to receptors on the dendrites of the next neuron:

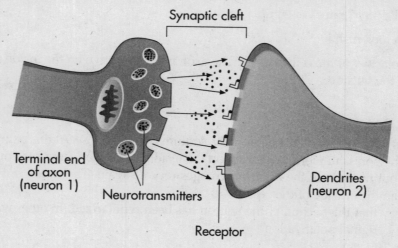

This usually triggers an action potential in the second neuron if the synaptic membrane is excited. Now, the impulse moves along the second neuron from dendrites to axon:

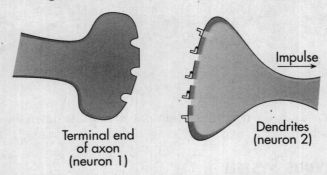

Terminal end
of axon
(neuron 1)

Impulse

Dendrites
(neuron 2)

There are many neurotransmitters, but the one we need to know for the AP Biology Exam is called **acetylcholine**. Acetylcholine is a neurotransmitter that:

- Is released from the end of an axon, when Ca^{2+} moves into the terminal end of the axon.

- Is picked up almost "instantly" by the dendrites of the next neuron.

- Can stimulate muscles to contract or inhibit postsynaptic potential.

Nerve impulses are transmitted across synapses by a number of neurotransmitters. It is released between neurons in the parasympathetic system, which we will discuss shortly.

The excess acetycholine in the synaptic cleft is broken down by the enzyme **acetylcholinesterase**. **Norepinephrine** is a peptide neurotransmitter that is released between neurons within the central nervous system. Another neurotransmitter, GABA, is secreted in the central nervous system and acts as an inhibitor.

Speed of an Impulse

Sometimes a neuron has supporting cells that wrap around its axon. These cells are called **Schwann cells**. Schwann cells produce a substance called the **myelin sheath**, which insulates the axon:

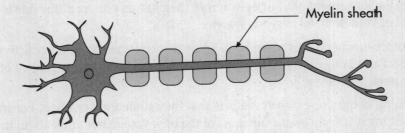

Myelin sheath

As you can see from the illustration above, the whole axon isn't covered with myelin sheaths. The spaces between myelin sheaths—the exposed regions of the axon—are called the **nodes of Ranviers**. Myelin sheaths speed up the propagation of an impulse. Instead of the standard "domino effect" that occurs during an action potential, the impulse can now jump from node to node. This form of conduction is called **saltatory conduction**.

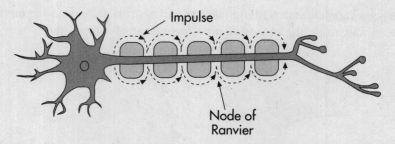

Impulse

Node of
Ranvier

Thanks to the myelin sheath, the neuron can transmit an impulse down the axon far more rapidly than it could without its help.

PARTS OF THE NERVOUS SYSTEM

The nervous system can be divided into two parts: the **central nervous system** and the **peripheral nervous system**.

Central Nervous System

All of the neurons within the brain and spinal cord make up the central nervous system. All of the other neurons lying outside the brain and the spinal cord—in our skin, our organs, and our blood vessels—are collectively part of the peripheral nervous system. Although both of these systems are really part of one system, we still use the terms "central" and "peripheral."

So keep them in mind:

- The *central nervous system* includes the neurons in the brain and spinal cord.

- The *peripheral nervous system* includes all the rest.

Peripheral Nervous System

The peripheral nervous system is further broken down into the **somatic nervous system** and the **autonomic nervous system**.

- The somatic nervous system is the part that controls voluntary activities. For example, the movement of your eyes across the page as you read this line is under the control of your somatic nervous system.

- The autonomic nervous system is that part that controls involuntary activities. Your heartbeat and your digestive system, for example, are under the control of the autonomic nervous system.

The interesting thing about these two systems is that they sometimes overlap. For instance, you can control your breathing if you choose to. Yet most of the time you do not think about it: Your "somatic" system hands control of your respiration over to the "autonomic" system.

The autonomic system is broken down even further to the **sympathetic nervous system** and the **parasympathetic nervous system**. The sympathetic system controls the **"fight-or-flight" response**.

When confronted with a threatening situation, an organism can choose to fight or to flee (thus fight or flight). To get ready for a quick, effective action, whether that be brawling or bolting, the two systems work antagonistically. The sympathetic nervous system raises your heart and respiration rates, causes your blood vessels to constrict, increases the levels of glucose in your blood, and produces "goosebumps" on the back of your neck. It even reroutes your blood sugar to your skeletal muscles in case you need to make a break for it. After the threat has passed the parasympathetic nervous system brings the body back to **homeostasis**—that is, back to normal. It lowers your heart and respiratory rates and decreases glucose levels in the blood.

The flow chart below will give you a nice overview of the different parts of the nervous system:

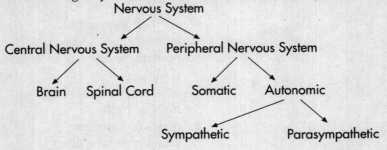

PARTS OF THE BRAIN

The brain can also be divided into parts. Here's a summary of the major divisions within the brain.

DIVISIONS WITHIN THE BRAIN	
Parts of the Brain	**Function**
Cerebrum	Controls all voluntary activities; receives and interprets sensory information
Cerebellum	Coordinates muscle activity
Hypothalamus	Regulates homeostasis and secretes hormones
Medulla	Controls involuntary actions such as breathing and swallowing

KEY WORDS

nerve net
ganglia
neurons
cell body
dendrites
axon
sensory neurons
motor effector neurons
interneurons
polarized
threshold
action potential
all-or-none response
depolarization
repolarized
sodium-potassium pump
refractory period
axon bulb
neurotransmitter
synapse
acetylcholine

acetylcholinesterase
epinephrine
norepinephrine
GABA
Schwann cells
myelin sheath
nodes of Ranviers
saltatory conduction
central nervous system
peripheral nervous system
somatic nervous system
autonomic nervous system
sympathetic nervous system
parasympathetic nervous system
fight-or-flight response
homeostasis
cerebrum
cerebellum
hypothalamus
medulla

VII: THE MUSCULOSKELETAL SYSTEM

Most organisms need some form of support. Many animals wear their support on the *outside*. They have an **exoskeleton**—a hard covering or shell. Insects, for example, have an exoskeleton made of chitin. All **vertebrates** (animals with backbones) possess an **endoskeleton**—their entire skeleton is on the *inside*. In addition to ourselves, fish, amphibians, reptiles, birds, and all other mammals are considered vertebrates and therefore have endoskeletons.

THE HUMAN SKELETAL SYSTEM

In humans, the supporting skeleton is made of **cartilage** and **bone**. Cartilage is found in the embryonic stages of all vertebrates. It is later replaced by bone, except in your external ear or the tip of your nose. Here's one thing you should remember: Bone is a connective tissue that contains nerves and blood vessels. Cartilage, on the other hand, lacks nerves and blood vessels.

Bones

Bone is made up of two substances: **collagen** and **calcium salts**. Bones are held together by **joints**, like the ball-and-socket joint in your shoulder. What holds the joints together? They're held together by tough connective tissues called **ligaments**. Just remember that ligaments attach bone to bone. Bones not only serve as support but together with muscles also help us move about. The connective tissues that attach muscles to bones are called **tendons**.

Muscles

There are three kinds of muscle tissue: **skeletal**, **smooth**, and **cardiac**. For the AP Biology Exam, you'll need to know the differences among the types of muscles.

Skeletal muscles control voluntary movements. You'll notice that they have stripes called **striations**. They are also **multinucleated**. Let's look at a more detailed level of a skeletal muscle under a microscope:

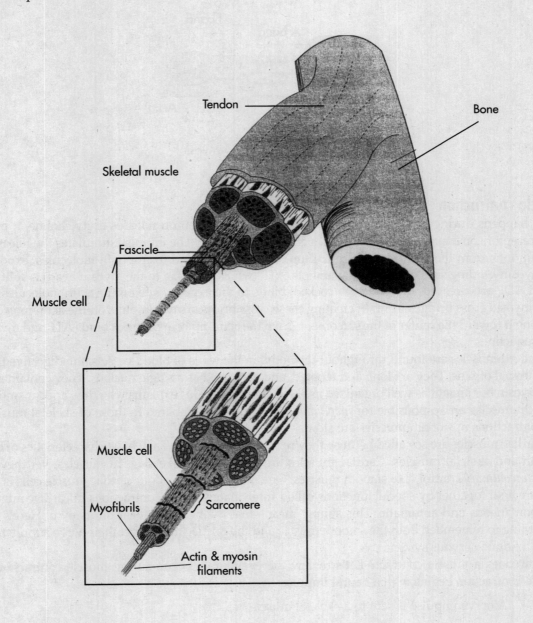

Organization of skeletal muscle

Muscles are made up of **muscle bundles**, which subdivide into **muscle fascicle**. Within each muscle fascicle are units called **muscle fiber cells**. Within each muscle fiber are contractile fibrils called **myofibrils**. A single myofibril is subdivided, by Z lines, into **sarcomeres** or contractile units.

The functional unit in a muscle cell is the **sarcomere**. Inside a sarcomere, there are two protein filaments: **actin** and **myosin**. Actins are the thin filaments, and myosin are the thick filaments:

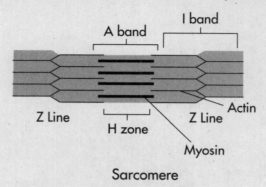

Sarcomere

Muscle contraction

What happens during a muscle contraction? When a motor neuron releases acetylcholine, it binds with receptors on a muscle fiber and causes an action potential. The impulse stimulates the release of calcium ions from the **sarcoplasmic reticulum**. Calcium ions bind to troponin molecules, exposing the myosin-binding sites on the actin filaments. ATP (which is bound to the myosin head) is split and P_i and ADP are released. Myosin, now cocked, binds to the exposed site on the actin molecules and actin-myosin cross bridges form. In creating these cross bridges, myosin pulls on the actin molecule, drawing it toward the center of the sarcomere. Then the actin-myosin complex binds ATP and myosin releases actin.

Smooth muscles are found throughout the body: in the walls of blood vessels, the digestive tract, and internal organs. They are long and tapered, and each cell has a single nucleus. They contain actin and myosin but are not as well organized as skeletal muscles. This explains why they appear smooth. Smooth muscles are responsible for *involuntary* movements. Compared to those of skeletal muscles, the contractions in smooth muscles are slow.

Cardiac muscles are so called because they're found in the heart. They have characteristics of both smooth and skeletal muscles. Cardiac muscles are striated, just like skeletal muscles, yet they are under *involuntary* control, like smooth muscles. One unique feature about cardiac muscle cells is that they are held together by special junctions called **intercalated discs**. Contractions in cardiac muscles are spontaneous and automatic. This simply means that the heart can beat on its own. Here's one more thing to remember: Both the smooth muscle and the cardiac muscle get their nerve supply from the autonomic nervous system.

How does a muscle contract? Let's review the events that occur during muscle contraction. A muscle contraction begins with a neural impulse:

- A nerve impulse is sent to a skeletal muscle.

- The neuron sending the impulse releases a neurotransmitter onto the muscle cell.

- The muscle depolarizes.

- Depolarization causes the sarcoplasmic reticulum to release calcium ions.

- These calcium ions cause the actin and myosin filaments to slide past each other.
- The muscle contracts.

Let's compare the types of muscle tissues:

TYPES OF MUSCLE TISSUES			
	Skeletal	**Smooth**	**Cariac**
Location	Attached to skeleton	Wall of digestive tract, inside the blood vessels	Wall of heart
Type of control	Voluntary	Involuntary	Involuntary
Striations	Yes	No	No
Multinucleated	Yes	No	No
Speed of contraction	Rapid	Slowest	Intermediate

KEY WORDS

exoskeleton
vertebrates
endoskeleton
cartilage
bone
collagen
calcium salts
joints
ligaments
tendons

skeletal muscles
smooth muscles
cardiac muscles
striations
sarcomere
multinucleate cells
actin
myosin
sarcoplasmic reticulum
intercalated discs

VIII: THE ENDOCRINE SYSTEM

Chemical messengers can be produced in one region of the body to act on target cells in another part. These chemicals, known as **hormones**, are produced in specialized organs called **endocrine glands**. An endocrine gland releases hormones directly into the bloodstream, which carries them throughout the body. Take a look at the endocrine glands in the human body:

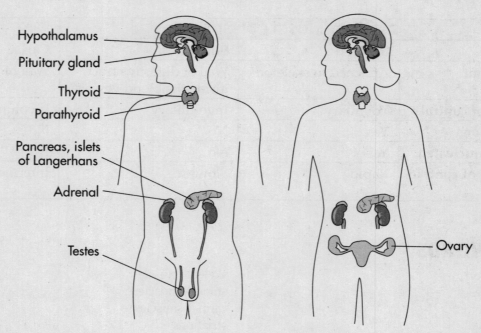

Before we launch into a review of the different hormones in the body, let's talk about how hormones work. Although hormones flow in your blood, they affect only specific cells. The cells that a hormone affects are known as the **target cells**. Suppose, for example, that gland X makes hormone Y. Hormone Y, in turn, has some effect on organ Z. We would then say that organ Z is the *target organ* of hormone Y.

Hormones also operate by a **negative feedback system**. That is, an excess of the hormone will signal the endocrine gland to temporarily shut down production. For example, when hormone Y reaches a peak level in the bloodstream, the organ secreting the hormone, gland X, will get a signal to stop producing hormone Y. Once the levels of hormone Y decline, the gland can resume production of the hormone.

THE PITUITARY

The pituitary is called the master gland because it releases many hormones that reach other glands and stimulate them to secrete their own hormones. The anterior pituitary therefore has many *target* organs. The pituitary has two parts: the **anterior pituitary** and the **posterior pituitary**. Each part secretes its own set of hormones.

The anterior pituitary secrets six hormones, three of which regulate growth and other organs. The other three are involved in regulating the reproductive systems. The hormones of the pituitary are:

- **Growth hormone (GH)**—stimulates growth throughout the body, targets bones and muscles.

- **Adrenocorticotropic hormone (ACTH)**—stimulates the adrenal cortex to secrete glucocorticoids and mineralocorticoids.

- **Thyroid-stimulating hormone (TSH)**—stimulates the thyroid to secrete thyroxine.

- **Follicle-stimulating hormone (FSH)**—stimulates the follicle to grow in females, and spermatogenesis in males.

- **Luteinizing hormone (LH)**—causes the release of the ovum during the menstrual cycle in females, and testosterone production in males.

- **Prolactin**—stimulates the mammary glands to produce milk.

The pituitary works in tandem with a part of the brain called the **hypothalamus**. The pituitary sits just below the hypothalamus:

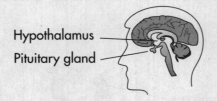

The hypothalamus regulates the anterior pituitary by secreting neurosecretory hormones that can stimulate or inhibit the actions of the anterior pituitary. The other part of the pituitary, the posterior pituitary, secretes two hormones:

- **Antidiuretic hormone** (or **vasopressin**)

- **Oxytocin**

These hormones are actually made in the hypothalamus but *stored* in the posterior pituitary. How about a mnemonic? When you think of the *pituitary*, think of the *GATOR pit*:

Growth hormone

ACTH

Thyroid-stimulating hormone

Oxytocin

R, for antidiu**R**etic hormone (or vasopressin)

(All of which come from the **pit**, or "**pituitary**" gland.)

And, to make things easier, the first three come first, or are "anterior" to the last two, which are "posterior." For us, that means that **G**, **A**, and **T** come from the **anterior pit**, while **O** and **R** come from the **posterior pit**.

As for the other three hormones, think of **FLAP**. FSH, LH, and prolactin are all hormones that have to do with the reproductive systems. Together, these two mnemonics should help you keep the different hormones of the pituitary gland straight. Now let's move on to the target organs.

The Pancreas

We already know that the pancreas produces enzymes that it releases into the small intestine via the pancreatic duct. The pancreas also secretes two hormones, **glucagon** and **insulin**, both of which are produced in clusters of cells called the **islet of Langerhans**. The target organs for these hormones are the liver and muscle cells. Glucagon, produced by α cells, stimulates the liver to convert **glycogen** into glucose and to release that glucose into the blood. Glucagon therefore *increases* the levels of glucose in the blood. Insulin has precisely the opposite effect that glucagon does.

When the blood has too much glucose floating around, insulin, produced by β cells, allows body cells to remove glucose from the blood. Consequently, insulin *decreases* the level of glucose in the blood. Insulin is particularly effective on muscle and liver cells. In short:

- *Insulin* lowers the blood sugar level.

- *Glucagon* raises the blood sugar level.

THE ADRENAL GLANDS

The adrenal glands contain two separate endocrine glands. One is called the **adrenal cortex** and the other is called the **adrenal medulla**. Although they are part of the same organ, these two endocrine glands have very different effects on the body. Let's start by discussing the adrenal cortex.

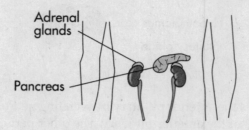

The adrenal cortex

Earlier we mentioned that **adrenocorticotropic hormone**, or simply **ACTH**, targets the adrenal cortex. When ACTH is released from the pituitary, it stimulates the adrenal cortex to produce and secrete its different hormones.

One group of hormones released by the adrenal cortex is the **glucocorticoids**. They increase the blood's concentration of glucose and help the body adapt to stress. In fact, glucocorticoids accomplish the same thing as glucagon, but in a slightly different way. Glucocorticoids promote the conversion of amino acids and fatty acids to glucose.

Another set of hormones is the **mineralocorticoids**. They help the body retain Na^+ and water in the kidneys. They accomplish this by promoting the reabsorption of sodium (Na^+) and chlorine (Cl^-), which get together to form common salt (NaCl). When salt is retained, water soon follows.

So then, just remember that the adrenal cortex releases two types of hormones:

- **Glucocorticoids**—target the liver and promote the release of glucose.

- **Mineralocorticoids**—target the kidney and promote the retention of water.

The adrenal medulla

The other adrenal gland, the adrenal medulla, is often referred to as the "emergency gland." It secretes two hormones: **epinephrine** and **norepinephrine**. These are the hormones involved in the fight-or-flight response. Both epinephrine and norepinephrine "kick in" under extreme stress. They increase your heart rate, metabolic rate, blood pressure, and give you a quick boost of energy.

THE THYROID

The thyroid gland, which is located in the neck, is the target organ of the thyroid-stimulating hormone (TSH):

Thyroid

When the thyroid is stimulated by TSH, it releases the hormone **thyroxine**. Thyroxine, which contains iodine, is responsible for regulating the metabolic rate in your body tissues. Two conditions are associated with thyroid hormones. **Hyperthyroidism** occurs in individuals who regularly release too much thyroxine. They have a fast metabolic rate and tend to be irritable and nervous. On the other hand, individuals who suffer from **hypothyroidism** have too little thyroxine circulating in their bloodstream. They exhibit a slow metabolic rate and tend to be sluggish and overweight. So let's summarize:

> An individual with *hypo*thyroidism has a slow metabolic rate, while an individual with *hyper*thyroidism has a fast metabolic rate.

The thyroid also secretes another hormone called **calcitonin**. This hormone decreases your blood's concentration of calcium by concentrating free-floating calcium in the bones. You'll recall from our discussion of bone cells that bones contain collagen and calcium salts. Calcitonin is responsible for depositing these calcium molecules in the bones.

THE PARATHYROIDS

The **parathyroids** are four little pea-shaped organs that rest on the thyroid. They secrete the hormone **parathyroid hormone**. Parathyroid hormone increases your blood calcium levels. Consequently, parathyroid hormone has the opposite effect that calcitonin does. If your blood needs more calcium, parathyroid hormone releases calcium ions stored in the bones. This process of building or breaking down bones to store and release calcium is called **bone remodeling**.

THE SEX HORMONES

Three hormones that we'll discuss in detail in the chapter on reproduction are **estrogen**, **progesterone**, and **testosterone**. Estrogen and progesterone are hormones released by the ovaries. Both hormones regulate the menstrual cycle. Testosterone is the male hormone responsible for promoting spermatogenesis, the production of sperm. In additon, these hormones maintain secondary sex characteristics.

How Hormones Work

How do hormones trigger the activities of their target cells? That all depends on whether the hormone is a **steroid** (lipid soluble) or a **protein** (not lipid soluble). If the hormone is a steroid, then the hormone can diffuse across the membrane of the target cell. It then binds to a receptor protein in the nucleus and activates specific genes contained in the DNA, which in turn make proteins.

However, if the hormone is a protein itself, it can't get into the target cell by means of simple diffusion. Remember: "Like dissolves like." The hormone binds to a receptor protein on the *cell membrane* of the target cell. This protein in turn stimulates the production of a second messenger called **cyclic AMP (cAMP)**. The cAMP molecule then triggers various enzymes, leading to specific cellular changes. A summary of the hormones and their effects on the body:

ORGAN	HORMONES	EFFECT
Anterior Pituitary	FSH	Stimulates activity in ovaries and testes
	LH	Stimulates activity in ovary (release of ovum) and production of testosterone
	ACTH	Stimulates the adrenal cortex
	Growth Hormone	Stimulates bone and muscle growth
	Prolactin	Causes milk secretion
Posterior Pituitary	Oxytocin	Causes uterus to contract
	Vasopressin	Causes kidney to reabsorb water
Thyroid	Thyroid Hormone Calcitonin	Regulates metabolic rate Lowers blood calcium levels
Parathyroid	Parathyroid Hormone	Increases blood calcium concentration
Adrenal Cortex	Aldosterone	Increases Na^+ and H_2O reabsorption in kidney
Adrenal Medulla	Epinephrine Norepinephrine	Increases blood glucose level and heart rate
Pancreas	Insulin	Decreases blood sugar concentration
	Glucagon	Increases blood sugar concentration
Ovaries	Estrogen	Promotes female secondary sex characteristics and thickens endometrial lining
	Progesterone	Maintains endometrial lining
Testes	Testosterone	Promotes male secondary sex characteristic and spermatogenesis

While the nervous system and the endocrine system work in close coordination, there are significant differences between the two:

- The nervous system sends nerve impulses using neurons, whereas the endocrine system secretes hormones.

- Nerve impulses control rapidly changing activities, such as muscle contractions, whereas hormones deal with long-term adjustments.

KEY WORDS

hormones
endocrine glands
target cells
negative feedback system
pituitary
anterior pituitary
posterior pituitary
growth hormone (GH)
thyroid stimulating hormone (TSH)
follicle-stimulating hormone (FSH)
luteinizing hormone (LH)
prolactin
hypothalamus
antidiuretic hormone (vasopressin)
oxytocin
glucagon
insulin
islet of Langerhans
glycogen
adrenal cortex

adrenal medulla
adrenocorticotropic hormone (ACTH)
glucocorticoids
mineralocorticoids
cortical sex hormones
epinephrine
norepinephrine
thyroxine
hyperthyroidism
hypothyroidism
calcitonin
parathyroids
parathyroid hormone
bone remodeling
estrogen
progesterone
testosterone
steroid
protein
cyclic AMP (cAMP)

IX: THE REPRODUCTIVE SYSTEM

Reproduction in animals involves the production of eggs and sperm.

Before we launch into reproduction, let's take a look at something that ties in very nicely with everything we've just discussed about hormones and the endocrine system: the menstrual cycle.

Since the ovaries release hormones, they are considered endocrine glands. The **ovaries** have two main responsibilities:

- They manufacture **ova.**

- They secrete **estrogen** and **progesterone**, sex hormones that are found in females.

The hormones secreted by the ovaries are involved in the menstrual cycle.

THE MENSTRUAL CYCLE

Phase 1: The follicular phase

In phase 1, the anterior pituitary secretes two hormones: **follicle-stimulating hormone** (FSH) and **luteinizing hormone** (LH). The FSH stimulates several follicles in the ovaries to grow. Eventually, one of these follicles gains the lead and dominates the others, which soon stop growing. The one growing follicle now takes command.

Because the follicle is growing in this phase, the phase itself is known as the **follicular phase**.

Remember that during all this time the follicle is releasing *estrogen*. Estrogen helps the uterine lining to grow and eventually causes the pituitary to release a very large amount of LH. This increase in estrogen causes a sudden surge in luteinizing hormone. This release of LH is known as a **luteal surge**. LH triggers **ovulation**—the release of the follicle from the ovary.

There are thus three hormones associated with the follicular phase:

- **Follicle-stimulating hormone** (FSH)—originates in the pituitary gland.

- **Estrogen**—originates in the follicle.

- **Luteinizing hormone** (LH)—originates in the pituitary gland.

The luteal surge makes the follicle burst and release the ovum. The ovum then begins its journey into the **fallopian tube**, which is also known as the **oviduct**. This is a crucial event in the female menstrual cycle and is known as **ovulation**. Once the ovum has been released, the follicular phase ends and the ovum is ready to move on to the next phase.

In addition to the growth of the follicle, the follicular phase involves the thickening of the **uterine walls**, or **endometrium**. This happens in preparation for the implantation of a fertilized cell. It stimulates secondary sex characteristics in girls. The entire follicular phase lasts about 10 days.

Phase 2: The luteal phase

By the end of the follicular phase, the ovum has moved into the fallopian tube and the follicle has been ruptured and left behind in the ovary. However, the ruptured follicle (now a fluid-filled sac) continues to function in the menstrual cycle. At this stage, it condenses into a little yellow blob called the **corpus luteum**, which is Latin for "yellow body."

The corpus luteum continues to secrete estrogen. In addition, it now starts producing the other major hormone involved in female reproduction, progesterone. Progesterone is responsible for readying the body for pregnancy. It does this by promoting the growth of glands and blood vessels in the endometrium. Without progesterone, a fertilized ovum cannot latch onto the uterus and develop into an embryo. We can therefore think of progesterone as the hormone of pregnancy.

After about 13 to 15 days, if fertilization and implantation have not occurred, the corpus luteum shuts down. Once it has stopped producing estrogen and progesterone, the final phase of the menstrual cycle begins.

Phase 3: The flow phase, or menstruation

Once the corpus luteum turns off, the uterus can no longer maintain its thickened walls. It starts to reabsorb most of the tissue that the progesterone encouraged it to grow. However, since there is too much to reabsorb, a certain amount is shed. This "sloughing off," or bleeding, is known as **menstruation**.

With the end of menstruation, the cycle starts all over again, readying the body for fertilization. Let's recap some of the major steps:

- In the follicular phase, the pituitary releases FSH, causing the follicle to grow.

- The follicle releases estrogen, which helps the endometrium to grow

- Estrogen causes the pituitary to release LH, resulting in a luteal surge.

- This excess LH causes the follicle to burst, releasing the ovum during ovulation.

- The shed follicle becomes the corpus luteum, which produces progesterone.

- Progesterone, the "pregnancy hormone," enhances the endometrium, causing it to thicken with glands and blood vessels.

- If fertilization does not occur after about two weeks, the corpus luteum dies, leading to menstruation—the sloughing off of uterine tissue.

If pregnancy occurs, the extraembryonic tissue of the fetus releases **human chorionic gonadotropin (HCG),** which helps maintain the uterine lining. In addition to the events themselves, the AP Biology Exam is likely to contain questions dealing with where the stages of the menstrual cycle and fertilization occur. Take a look at the following diagram. Familiarize yourself with the parts of the female reproductive system and pay special attention to the different sites of the stages we've just discussed:

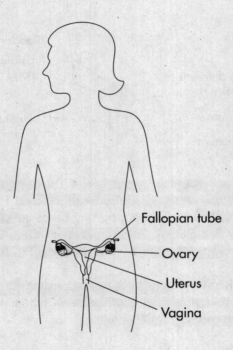

Remember:

- The follicles (and thus the ova) are contained in the ovaries.
- Hormones are released from the ovaries and pituitary gland.
- Fertilization occurs in the fallopian tubes.
- The fertilized ovum implants itself in the uterus.

THE MALE REPRODUCTIVE SYSTEM

Now let's discuss the hormones in the male reproductive system. **Testosterone,** along with the cortical sex hormones we saw earlier, is responsible for the development of the sex organs and secondary sex characteristics. In addition to the deepening of the voice, these characteristics include body hair, muscle growth, and facial hair, all of which indicate the onset of **puberty**. Testosterone also has another function. It stimulates the testes, the male reproductive organs, to manufacture **sperm cells**.

Testosterone does this by causing cells in the testes to start undergoing meiosis.

Take a look at the male reproductive system:

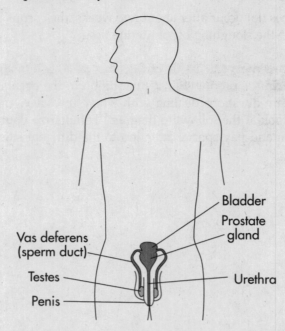

Sperm and male hormones are produced in the testes. The main tissues of the testes, called the **seminiferous tubules**, are where spermatogonia undergo meiosis. The spermatids then mature in the epididymis. The interstitial cells, which are supporting tissue, produce testosterone. Sperms then travel through the **vas deferens** and pick up fluids from the **seminal vesicles** (which provides them with fructose for energy) and the **prostate gland** (which provides an alkaline solution that neutralizes the vagina's acidic fluids).

Unlike the female reproductive system, male hormones are continually secreted throughout the life of the male. **FSH** targets the seminiferous tubules of the testes, where it stimulates sperm production. **LH** stimulates interstitial cells to produce testosterone.

KEY WORDS

ovaries

ova

estrogen

progesterone

follicle-stimulating hormone (FSH)

luteinizing hormone (LH)

follicular phase

luteal surge

fallopian tube

oviduct

ovulation

uterine walls

endometrium

corpus luteum

menstruation

testosterone

puberty

sperm cells

X: EMBRYONIC DEVELOPMENT

How does a tiny, single-celled egg develop into a complex, multicellular organism? By dividing, of course. The cell will change shape and organization many times by going through a succession of stages. This process is called **morphogenesis**. In order for the human sperm to fertilize an egg it must dissolve the **corona radiata**, a dense covering of follicle cells that surrounds the egg. Then the sperm must penetrate the **zona pellucida**, the zone below the corona radiata.

When an egg is fertilized by a sperm it forms a diploid cell called a **zygote**:

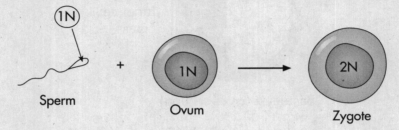

Sperm   +   Ovum   Zygote

Fertilization triggers the zygote to go through a series of rapid cell divisions called **cleavage**. What's interesting at this stage is that the embryo doesn't grow. The cells just keep dividing to form a solid ball called a **morula**:

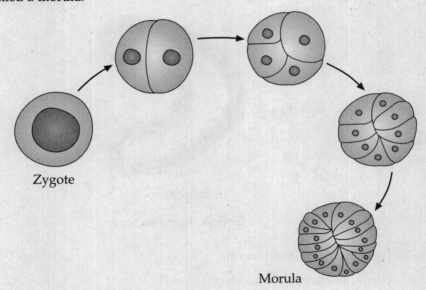

Zygote

Morula

One cell becomes two cells, two cells become four cells, etc.

BLASTULA

The next stage is called **blastula**. As the cells continue to divide, they "press" against each other and produce a fluid-filled cavity called a **blastocoel**:

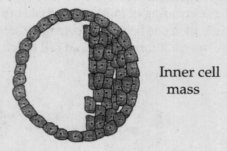

Inner cell mass

Bastocyst (cross section)

GASTRULA

During **gastrulation**, the zygote begins to change its shape. Cells now migrate into the blastocoel and differentiate to form three germ layers: the **ectoderm**, **mesoderm**, and **endoderm**:

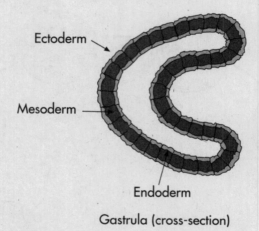

Ectoderm

Mesoderm

Endoderm

Gastrula (cross-section)

- The outer layer becomes the ectoderm.
- The middle layer becomes the mesoderm.
- The inner layer becomes the endoderm.

Each germ layer gives rise to various organs and systems in the body. Here's a list of the organs that develop from each germ layer.

- The **ectoderm** produces the epidermis (the skin), the eyes, and the nervous system.
- The **endoderm** produces the inner linings of the digestive tract and respiratory tract, as well as accessory organs such as the pancreas, gall bladder, and liver. They are called "accessory" organs because they are offshoots of the digestive tract, as opposed to the channels of the tract itself.
- The **mesoderm** gives rise to everything else. This includes bones and muscles as well as the excretory, circulatory, and reproductive systems.

NEURULA

The **neurula** stage begins with the formation of two structures; the **notochord,** a rod-shaped structure running beneath the nerve cord, and the **neural tube** cells, which develop into the central nervous system. By the end of this stage, we're well on our way to developing a nervous system:

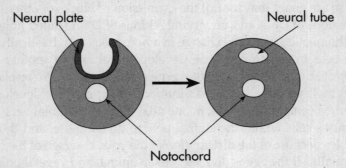

Neural plate

Neural tube

Notochord

As far as the AP Biology Exam is concerned, the *order* of the stages and the various events is extremely important. For our purposes, think of the order of embryological development in this way:

Zygote → Cleavage → Blastula → Gastrula → Neurula

WHAT ABOUT CHICKEN EMBRYOS?

In addition to the primary germ layers, some animals have **extraembryonic membranes**. Your average developing chicken, for example, possesses these membranes.

There are basically four extraembryonic membranes: the **yolk sac**, **amnion**, **chorion**, and **allantois**. These extra membranes are common in birds and reptiles.

For the AP Biology Exam, you should be familiar with these membranes and their functions:

FUNCTIONS OF EXTRAEMBRYONIC MEMBRANES	
Extraembryonic membrane	**Function**
Yolk sac	Provides food for the embryo
Amnion	Forms a fluid-filled sac that protects the embryo
Allantois	Membrane involved in gas exchange
Chorion	Otermose membrane that surronds all the other extramebryonic membranes

Fetal Embryo

The fetal embryo also has extraembryonic membranes during development: the amnion, chorion, allantois, and yolk sac. The placenta and the umbilical cord are outgrowths of these membranes. The placenta is the organ that provides the fetus with nutrients and oxygen and gets rid of the fetus's wastes. The placenta develops from both the chorion and the uterine tissue of the mother. The umbilical cord is the organ that connects the embryo to the placenta.

DEVELOPMENT OF AN EMBRYO

Induction: The ability of one tissue to determine the fate of another tissue is called induction. Certain cells, called **organizers**, release a chemical substance (**a morphogen**) that moves from one tissue to the target tissue. It is now known that development involves many episodes of embryonic induction.

Homeotic genes: These are genes that control the expression of blocks of other genes. The homeotic genes have variable regions of DNA and unvarying regions of DNA (called **homeobox**). Since every organism has several homeotic genes, any change in a homeobox can drastically affect development. Mutations that cause cells to switch from one developmental fate to another are called **homeotic** mutations. These earliest studies about homeotic genes were done on *Drosophila*.

Cytoplasm: The cytoplasm can also have an influence on embryonic development. For example, chicken and frog embryos contain more yolk in one pole (the vegetal pole) versus the other pole (the animal pole). This causes cells within the animal pole to divide more, and thus to be smaller than those of the vegetal pole. Because of the distribution of the yolk, cleavage of the egg does not produce eggs that develop normally. If the egg is divided into an animal and vegetal pole, development does not proceed normally.

KEY WORDS

morphogenesis
zygote
fertilization
cleavage
morula
blastula
blastocoel
gastrulation
ectoderm
mesoderm

endoderm
neurula
notochord
neural tube
extraembryonic membranes
yolk sac
amnion
chorion
allantois

QUIZ

Directions: Each of the questions or incomplete statements below is followed by five suggested answers or completions. Select the one that is best in each case. Answers can be found on pages 243–245.

1. The blood type that is the universal recipient is

   (A) O
   (B) A
   (C) AB
   (D) B
   (E) Rh

2. Special structures for absorption of fats in the small intestine are called

   (A) nephrons
   (B) lacteals
   (C) villi
   (D) root hairs
   (E) hormones

3. The concentration of which of the following hormones in the bloodstream stimulates the contraction of uterine muscles in females?

   (A) FSH
   (B) LH
   (C) Prolactin
   (D) Oxytocin
   (E) Epinephrine

4. All of the following are directly involved in immune responses in mammals EXCEPT:

   (A) the thymus gland
   (B) the bone marrow
   (C) the lymphatic system
   (D) the spleen
   (E) the kidney

5. Compared to skeletal muscles, smooth muscle cells are

   (A) uninucleated, rapidly contracting, and under voluntary control
   (B) uninucleated, slowly contracting, and under involuntary control
   (C) multinucleated, rapidly contracting, and under voluntary control
   (D) uninucleated, slowly contracting, and under voluntary control
   (E) multinucleated, rapidly contracting, and contain intercalated discs

6. The enzyme that initially breaks down proteins in the human digestive system is called

   (A) bile
   (B) pepsin
   (C) trypsin
   (D) salivary amylase
   (E) pancreatic amylase

7. Which of the following is true regarding the hypothalamus?

   (A) It secretes thyroid-stimulating hormone.
   (B) It secretes luteinizing hormone.
   (C) It is an extension of the pituitary gland.
   (D) It produces neurosecretory hormones.
   (E) It secretes neurotransmitters into synapses.

8. The digestive enzyme that hydrolyzes molecules of fats into fatty acids is known as

   (A) bile
   (B) lipase
   (C) amylase
   (D) protease
   (E) trypsin

9. Diabetes mellitus is characterized by high blood glucose levels and can be caused by a shortage or absence of which of the following hormones?

   (A) Glucagon
   (B) Insulin
   (C) Parathyroid hormone
   (D) Calcitonin
   (E) Norepinephrine

10. All of the following are associated with the distribution of CO_2 in the bloodstream EXCEPT:

    (A) water
    (B) hemoglobin
    (C) plasma
    (D) red blood cells
    (E) platelets

11. Which of the following extraembryonic membranes stores waste products?

    (A) chorion
    (B) amnion
    (C) allantois
    (D) yolk sac
    (E) eggshell

Directions: Each group of questions consists of five lettered headings followed by a list of numbered phrases or sentences. For each numbered phrase or sentence, select the one heading that is most closely related to it and fill in the corresponding oval on the answer sheet. Each heading may be used once, more than once, or not at all in each group.

Questions 12–14

   (A) Right atrium
   (B) Left atrium
   (C) Right ventricle
   (D) Left ventricle
   (E) Vena cava

12. Sends blood to the lungs

13. Receives blood from the body

14. Sends blood from the head to the heart

Questions 15–18

    (A) Zygote
    (B) Blastocoel
    (C) Morula
    (D) Blastula
    (E) Gamete

15. Germ line cells in both males and females

16. A fluid-filled cavity that forms after several rounds of cell division

17. Results immediately after fertilization of the ovum by a sperm

18. A solid mass of cells produced by cleavage of the zygote.

Questions 19–22

    (A) Liver
    (B) Small intestine
    (C) Gall bladder
    (D) Pancreatic duct
    (E) Large intestine

19. Carries digestive enzymes to the small intestine

20. A storage organ for bile

21. Completely digests foodstuffs to simple substances

22. Reabsorbs water into the bloodstream

Questions 23–26

    (A) Cartilage
    (B) Tendons
    (C) Ligaments
    (D) Collagen
    (E) Bone

23. Connects bone to bone

24. Connects muscle to bone

25. Embryonic connective tissue found early in life

26. A mineralized connective tissue

Evolution

All of the organisms we see today arose from earlier organisms. This process, known as **evolution**, can be described as a change in a population over time. Interestingly, however, the driving force of evolution, **natural selection**, operates on the level of the individual. In other words, evolution is defined in terms of populations but occurs in terms of individuals.

NATURAL SELECTION

What is the basis of our knowledge of evolution? Much of what we now know about evolution is based on the work of **Charles Darwin**. Darwin was a 19th-century British naturalist who sailed the world in a ship named the HMS *Beagle*.

Darwin developed his theory of evolution based on natural selection after studying animals in the Galapagos Islands.

Darwin concluded that it was impossible for the finches and turtles of the Galapogos to simply "grow" longer beaks or necks. Rather, the driving force of evolution must have been natural selection. Quite simply put, this means that nature would "choose" which organisms survive on the basis of their fitness. For example, on the first island Darwin studied, there must once have been short-necked turtles. Unable to reach the higher vegetation, these turtles eventually died off, leaving only those turtles with longer necks. Consequently, evolution has come to be thought of as "the survival of the fittest": Only those organisms most fit to survive will survive.

Darwin elaborated his theory in a book entitled *On the Origin of Species*. In a nutshell, here's what Darwin observed:

- Each species produces more offspring than can survive.

- These offspring compete with one another for the limited resources available to them.

- Organisms in every population vary.

- The offspring with the most favorable traits or variations are the most likely to survive and therefore produce more offspring.

LAMARCK AND THE LONG NECKS

Darwin was not the first to propose a theory explaining the variety of life on earth. One of the most widely accepted theories of evolution in Darwin's day was that proposed by **Jean-Baptiste de Lamarck**.

In the 18th century, Lamarck had proposed that *acquired* traits were inherited and passed on to offspring. For example, in the case of our turtles, Lamarck's theory said that the turtles had long necks because they were constantly reaching for higher leaves while feeding. This theory is referred to as the "law of use and disuse," or, as we might say now, "use it or lose it." According to Lamarck, turtles have long necks because they constantly *use* them.

We know now that Lamarck's theory was wrong: Acquired changes—that is, changes at a "macro" level in somatic (body) cells—cannot be passed on to germ cells. For example, if you were to lose one of your fingers, your children would not inherit this trait. But how did Darwin, who lacked the understanding of genetics that we have today, come to dismiss Lamarck? What led him to this idea of natural selection?

In essence, nature "selects" which living things survive and reproduce. Today, we find support for the theory of evolution in several areas:

- **Paleontology**, or the study of fossils. Paleontology has revealed to us both the great variety of organisms (most of which, including trilobites, dinosaurs, and the woolly mammoth, have died off) and the major lines of evolution.

- **Biogeography**, or the study of the distribution of flora (plants) and fauna (animals) in the environment. Scientists have found related species in widely separated regions of the world. For example, Darwin observed that animals in the Galapagos have traits similar to those of animals on the mainland of South America. One possible explanation for these similarities is a common ancestor. As we'll see below, there are other explanations for similar traits. However, when organisms share multiple traits, it's pretty safe to say that they also shared a common ancestor.

- **Embryology**, or the study of the development of an organism. If you look at the early stages in vertebrate development, all the embryos look alike! All vertebrates—including fish, amphibians, birds, and even humans—show fish-like features called gill slits.

- **Comparative anatomy**, or the study of the anatomy of various animals. Scientists have discovered that some animals have similar structures that serve different functions. For example, a human's arm, a dog's leg, a bird's wing, and a whale's fin are all the same appendages, though they have evolved to serve different purposes. These structures, called **homologous structures**, also point to a common ancestor.

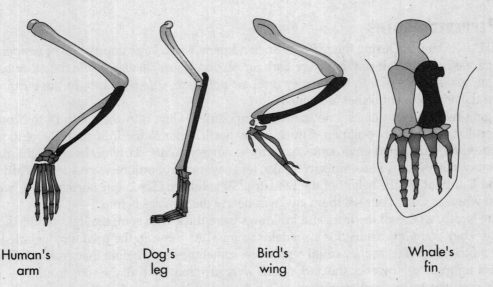

Human's arm

Dog's leg

Bird's wing

Whale's fin

In contrast, sometimes animals have features with the same function but that are structurally different. A bat's wing and an insect's wing, for example, are both used to fly. They therefore have the same function, but have evolved totally independently of one another. These are called **analogous structures**. Another classic example of an analogous structure is the eye. Though scallops, insects, and humans all have eyes, these three different types of eyes are thought to have evolved entirely independently of one another. They are therefore analogous structures.

- **Molecular biology**. Perhaps the most compelling proof of all is the similarity at the molecular level. Today, scientists can examine the nucleotide and amino acid sequences of different organisms. From these analyses, we've discovered that organisms that are closely related have a greater proportion of sequences in common than distantly related species. For example, most of us don't look much like chimpanzees. However, by some estimates, as much as 99% of our genetic code is identical to that of a chimp.

GENETIC VARIABILITY

In Chapter 8, we saw how traits are passed from parents to offspring. You'll recall from our discussion of heredity that different alleles are passed from parents to their progeny. For example, you might have an allele for brown eyes from your mother and an allele for blue eyes from your father. Since brown is dominant, you'll wind up with brown eyes. We also saw how these alleles are in fact just different forms of the same gene.

As you know, no two individuals are identical. The differences in each person are known as **genetic variability**. All this means is that no two individuals in a population have identical sets of alleles (except, of course, identical twins). How did all this wonderful variation come about? Through **random mutation**.

It might be hard to think of it in this way, but this is the very foundation of evolution, as we'll soon see. Now that we've reintroduced genes, we can refine our definition of evolution. More specifically:

Evolution is the change in the gene pool of a population over time.

THE PEPPERED MOTHS

Let's take an example. During the 1850s in England there was a large population of peppered moths. In most areas, exactly half of them were dark, or carried "dark" alleles, while the other half carried "light" alleles. All was fine in these cities until air pollution, due primarily to the burning of coal, changed the environment. What happened?

Imagine two different cities: one that was unpolluted, City 1, in the south of the country, and the other that was heavily polluted, City 2, in the north. Prior to the Industrial Revolution, both of these cities had unpolluted environments. In these environments, dark moths and light moths lived comfortably side by side. For simplicity's sake, let's say our proportions were a perfect fifty-fifty, half dark and half light. At the height of the Industrial Revolution, City 2, our northern city, was heavily polluted whereas City 1, our southern city, was nearly the same as before.

In the North, where all the trees and buildings were thick with soot, the light moths didn't stand a chance. They were impossible for a predator to miss! As a result, the predators gobbled up light-colored moths just as fast as they could reproduce, sometimes even before they reached an age where they *could* reproduce. However, the dark moths were just fine. With all the soot around the predators couldn't even see them; they continued doing their thing—above all, *reproducing*. And when they reproduced, they had more and more offspring carrying the dark allele.

After a few generations, the peppered moth gene pool in City 2 changed. Although our original moth gene pool was 50 percent light and 50 percent dark, excessive predation changed that. By about 1950, our gene pool reached 90 percent dark and only 10 percent light. This occurred because the light moth didn't stand a chance in an environment where it was so easy to spot. The dark moths, on the other hand, multiplied just as fast as they could.

In the southern city, you'll remember, there was very little pollution. What happened there? Things remained pretty much the same. The gene pool was unchanged, and even today you'll find roughly equal proportions of light moths and dark moths.

CAUSES OF EVOLUTION

The allele frequency remains constant in a population unless something happens to alter the gene pool. In the case above, the pollution in City 2 altered the frequency of certain alleles in the peppered moth population.

Individuals in a population are always competing. They compete for food, water, shelter, mates, etc. When a population is subjected to environmental change, or "stress," those who are better equipped to compete are more likely to survive. As we saw above, this process is called *natural selection*. In other words, nature "chooses" those members of a population best suited to survive. These survivors then have offspring who will carry many of the alleles that their parents carried, making it more likely that they, too, will survive.

As time goes on, more and more members of the population resemble the better competitors, while fewer resemble the poorer competitors. Over time, this will change the gene pool. The result is evolution.

Wait a second—didn't we say earlier that *random mutation* was the foundation of evolution? In fact, we did. Yet our example seems to say that the **environment** caused evolution. The truth lies somewhere between these two.

For natural selection to operate, there must be variation in a population. In this case, the variation was due to a mutation. By the way, what other possible sources of variation exist in a population? Sexual reproduction is a major factor in genetic variability. Meiosis, through crossing-over, produces new genetic combinations, as does the union of different haploid gametes to produce a diploid zygote.

Now let's go back to our moths. Why did the dark moths in the north of the country survive? Because they were dark-colored. But how did they become dark-colored? The answer is, through random mutation. One day, a moth was born with a dark-colored shell. As long as a mutation does not kill an organism (most mutations, in fact, do), it will be kept. Over time, this one moth had off-spring. These, too, were dark. The dark- and the light-colored moths lived happily side by side until something from the outside—in our example, the environment—changed all that.

The initial variation came about by chance. This variation gave the dark moths an edge. However, the edge did not become apparent until something made it apparent. In our case, that something was the intensive pollution due to the burning of coal. The abundance of soot made it easier for predators to spot the light-colored moths, thus effectively removing them from the population.

Eventually, over long stretches of time, these two different populations might change so much that they could no longer reproduce together. At that point, we would have two different species and we could say, definitively, that the moths had evolved. As a consequence of random mutation and the pressure put on the population by an environmental change, evolution occurred.

TYPES OF SELECTION

The situation with our moths is an example of **directional selection**. One of the phenotypes was favored at one of the extremes of the normal distribution.

In other words, directional selection "weeds out" one of the phenotypes. In our case, dark moths were favored and light moths were practically eliminated. Here's one more thing to remember: Directional selection can happen only if the appropriate allele—the one that is favored under the new circumstances—is already present in the population.

Two other types of selection are **stabilizing selection** and **disruptive selection**. Stabilizing selection means that organisms in a population with extreme traits are eliminated.

This type of selection favors organisms with common traits. It "weeds out" the phenotypes that are less adaptive to the environment. A good example is birth weight in human babies. If babies are abnormally small or abnormally large, they have a low rate of survival. The highest rate of survival is found among babies with an average weight. Disruptive selection, on the other hand, does the reverse. It favors both the extremes and selects against common traits. For example, females are "selected" to be small and males are "selected" to be large in elephant seals. You'll rarely find a female or male of intermediate size. Artificial selection, on the other hand, refers to the process by which a breeder *chooses* which traits to favor. A good example is how farmers breed seedless grapes.

SPECIES

A dog and a bumblebee obviously cannot come together to produce offspring. They are therefore different **species**. However, a poodle and a Great Dane could reproduce (at least in theory). We would not say that they are different species; they are merely different breeds.

Let's get back to our moths. We said above that evolution occurred when they could no longer reproduce. In fact, this is simply the endpoint of that particular cycle of evolution: **speciation**. Speciation refers to the emergence of new species. The type of evolution that our peppered moths underwent is known as **divergent evolution**.

Divergent evolution results in closely related species with different behaviors and traits. As with our example, these species often originate from a common ancestor. More often than not, the "engine" of evolution is cataclysmic environmental change, such as pollution in the case of the moths. Geographical barriers, new stresses, disease, and dwindling resources are all factors in the process of evolution.

Convergent evolution is the process in which two unrelated and dissimilar species come to have similar traits (analogous) often because they have been exposed to similar selective pressures. Examples of convergent evolution include aardvarks, anteaters, and pangolins. They all have strong, sharp claws and long snouts with sticky tongues to catch insects, yet they evolved from three completely different mammals.

There are two types of speciation: **allopatric speciation** and **sympatric speciation**. Allopatric speciation simply means that a population becomes separated from the rest of the species by a geographical barrier so that they can't interbreed. An example would be a mountain that separates two populations of ants. In time, the two populations might evolve into different species. If, however, new species form without any geographic barrier, it is called sympatric speciation. This type of speciation is common in plants. Two populations of plants may evolve in the same area without any geographic barrier.

POPULATION GENETICS

Mendel's laws can also extend to the population level. Suppose you caught a bunch of fruit flies—about 1,000. Let's say that 910 of them were red-eyed and 90 were green-eyed. If you allowed the fruit flies to mate and counted the next generation, we'd see that the ratio of red-eyed to green-eyed fruit flies would remain the same: 91% red-eyed and 9% green-eyed. That is, the allele frequency would remain constant. At first glance you may ask, how could that happen?

The **Hardy-Weinberg law** states that even with all the shuffling of genes that goes on, the relative frequencies of genotypes in a population still prevail over time. The alleles don't get lost in the shuffle. The dominant gene doesn't become more prevalent and the recessive gene doesn't disappear.

Let's say that the allele for red eyes, R, is dominant over the allele for green eyes, r. Red-eyed fruit flies include homozygous dominants, RR, and heterozygous, Rr. The green-eyed fruit flies are recessive, rr.

HARDY-WEINBERG EQUATIONS

The frequency of each allele is described in the equation below. The allele must be either R or r. Let "p" represent the frequency of the R allele and "q" represent the frequency of the other allele in the population.

$$p + q = 1$$

This sum of the frequencies must add up to one. If you know the value of one of the alleles, then you'll also know the value of the other allele.

We can also determine the frequency of the *genotypes* in a population using another equation:

$$p^2 + 2pq + q^2 = 1$$

In this equation, p^2 represents the homozygous dominants, $2pq$ represents the heterozygotes and q^2 represents the homozygous recessives.

So how do we use these equations? Use the proportions in the population to figure out both the allele and genotype frequencies. Let's calculate the frequency of the genotype for green-eyed fruit flies. If 9 percent of the fruit flies are green-eyed, then the *genotype* frequency, q^2, is 0.09. You can now use this value to figure out the frequency of the recessive allele in the population. The allele frequency for green eyes is equal to the square root of 0.09— that's 0.3. If the recessive allele is 0.3, the dominant allele must be 0.7. That's because 0.3 + 0.7 equals 1.

Using the second equation, you can calculate the genotypes of the homozygous dominants and the heterozygotes. The frequency for the homozygous dominants, p^2, is 0.7×0.7, which equals 0.49. The frequency for the heterozygotes, $2pq$, is $2 \times 0.3 \times 0.7$, which equals 0.42. If you include the frequency of the recessive genotype—0.09—the numbers once again add up to 1.

Hardy-Weinberg Equilibrium

The **Hardy-Weinberg law** says that a population will be in genetic equilibrium only if it meets these five conditions: 1) a large population, 2) no mutations, 3) no immigration or emigration, 4) random mating, and 5) no natural selection.

Violations of the Hardy-Weinberg Law

When these five conditions are met the gene pool in a population is pretty stable. Any departures from the Hardy-Weinberg law results in changes in allele frequencies in a population. For example, if a small group of your fruit flies moved to a new location, the allele frequency may be altered and result in evolutionary changes. That's an example of **genetic drift** called the founder effect. In other words, the gene frequency may differ from the original gene pool. Genetic drift often occurs in new colonies.

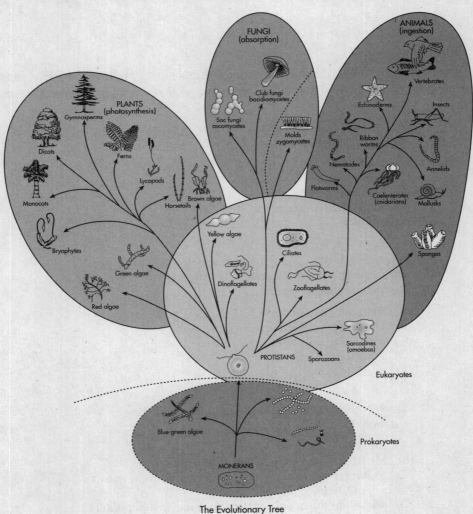

The Evolutionary Tree

KEY WORDS

evolution
natural selection
Charles Darwin
Jean-Baptiste de Lamarck
paleontology
biogeography
embryology
comparative anatomy
homologous structures
analogous structures
molecular biology
genetic variability
random mutation
alleles

environment
directional selection
stabilizing selection
disruptive selection
species
speciation
divergent evolution
allopatric speciation
sympatric speciation
Hardy-Weinberg law
genetic drift
convergent evolution
analagous structures

QUIZ

Directions: Each of the questions or incomplete statements below is followed by five suggested answers or completions. Select the one that is best in each case. Answers can be found on page 245.

1. The greatest degree of genetic variability would be expected among organisms that reproduce via

   (A) budding
   (B) sporulation
   (C) sexual recombination
   (D) vegetative propagation
   (E) mitosis

2. Kangaroos, koalas, and opossums are all marsupials, or animals with a maternal pouch. The diversity of marsupials in Australia is an example of

   (A) divergent evolution
   (B) convergent evolution
   (C) disruptive selection
   (D) genetic variability
   (E) speciation

3. Which of the following statements is true regarding mutations?

   (A) Mutations are irreversible.
   (B) Mutations are generally lethal to populations.
   (C) Mutations serve as a source for genetic variation in a
        population.
   (D) Mutations affect only certain gene loci in a population.
   (E) Mutation rates are very high in most populations.

Directions: Each group of questions consists of five lettered headings followed by a list of numbered phrases or sentences. For each numbered phrase or sentence, select the one heading that is most closely related to it and fill in the corresponding oval on the answer sheet. Each heading may be used once, more than once, or not at all in each group.

Questions 4–8

(A) Stabilizing selection
(B) Directional selection
(C) Sympatric speciation
(D) Allopatric speciation
(E) Disruptive selection

4. Mortality in an annual plant is highest among the extreme variants of that plant

5. Favors selection of both larger and smaller snails relative to intermediate variants

6. Favors selection of individuals with intermediate phenotypes over individuals with extreme phenotypes

7. Favors selection of organisms with longer limbs than the present average

8. A population that is geographically isolated from other members of its species gives rise to a separate species

Questions 9–12

(A) Use and disuse
(B) Natural selection
(C) Comparative anatomy
(D) Biogeography
(E) Paleontology

9. A body builder develops large muscles

10. The appearance of strains of bacteria that are no longer affected by certain antibiotics

11. The study of the past and present distribution of species

12. The fossil of an ancient horse, Hyracotherium, reveals two splints on either side of the large toe, indicating that it was a progenitor of the modern horse

13

Animal Behavior and Ecology

In the preceding chapters, we looked at individual organisms and the ways in which they solve life's many problems: acquiring nutrition, reproducing, etc. Now let's turn to how organisms deal with their environments. We can divide the discussion of organisms and their environments into two general categories: **behavior and ecology**.

BEHAVIOR

Some animals behave in a programmed way to specific stimuli while others behave according to some type of learning. We humans can do both. For the AP Biology Exam, you'll have to know a little about the different types of **behavior**.

All behavior means, basically, is how organisms cope with their environments. This chapter looks at the general types of behavior: instinct, imprinting, classical conditioning, operant conditioning, and insight. Let's start with instinct.

INSTINCTS

Instinct is an inborn, unlearned behavior. Sometimes the instinctive behavior is triggered by environmental signals called releasers. The releaser is usually a small part of the environment that is perceived. For example, when a male European robin sees another male robin, the sight of a tuft of red feathers on the male is a releaser that triggers fighting behavior. In fact, since instinct underlies all other behavior, it can be thought of as the circuitry that guides behavior.

For example, hive insects such as bees and termites never learn their roles; they are born knowing them. On the basis of this inborn knowledge, or instinct, they carry out all their other behaviors: A worker carries out "worker tasks," a drone "drone tasks," and a queen "queen tasks." Another example is the "dance" of the honey bee, which is used to communicate the location of food to other members of the beehive.

There are other types of instinct that last for only a part of an animal's life and are gradually replaced by "learned" behavior. For example, human infants are born with an ability to suck from a nipple. If it were not for this instinctual behavior, the infant would starve. Ultimately, however, the infant will move beyond this instinct and learn to feed itself. What exactly, then, is instinct?

For our purposes:

> Instinct is the inherited "circuitry" that directs and guides behavior.

A particular type of innate behavior is a **fixed action pattern**. These behaviors are not simple reflexes and yet they are not conscious decisions. An example is the egg-rolling behavior exhibited by a graylag goose. If the egg is removed from the goose, it will continue to make the same movements. That is, the innate movements are independent of the environment.

LEARNING

Another form of behavior is **learning**. Learning refers to a change in a behavior brought about by an experience (which is what you're doing this very moment). Animals learn in a number of ways. For the AP Biology Exam, we'll take a look at **imprinting, classical conditioning, operant conditioning,** and **insight**.

Imprinting

Have you ever seen a group of goslings waddling along after their mother? How is it that they recognize her? Well, the mother arrives and gives out a call that the goslings "recognize." The goslings, hearing the call, know that this is their mother, and follow her around until they are big enough to head out on their own.

Now imagine the same goslings, newly hatched. If the mother is absent, they will accept the first moving object they see as their mother. This process is known as **imprinting**.

Animals undergo imprinting within a few days after birth in order to recognize members of their own species. While there are different types of imprinting—including parent, sexual, and song imprinting—they all occur during a **critical period**—a window of time when the animal is sensitive to certain aspects of the environment.

Remember that:

> Imprinting is a form of learning that occurs during a brief period of time, usually early in an organism's life.

Classical conditioning

If you have a dog or a cat, you know that every time you hit the electric can opener, your cat or dog comes running. This is a form of **classical conditioning**. To feed your pet, you need to open its can of food. For your pet, the sound of the opener has come to be associated with eating: Every time it hears the opener, it thinks that it is about to be fed. We can say that your pet has been "conditioned" to link the buzz of the can opener and its evening meal.

The classic experiment demonstrating conditioning was done by a Russian scientist named Ivan Pavlov, who made his dogs salivate by ringing a bell. He did this the same way you make your dog come running with the can opener. Each time he fed his dogs, Pavlov rang a bell. Eventually, the dogs came to "associate" the bell with the food. By the end of the experiment, Pavlov had merely to ring the bell to start the dogs salivating. This type of learning is now known as **associative learning**.

Pavlov's experiments demonstrated what many of us know already: We can learn through conditioning, or repeated instances of an event.

Operant conditioning

Another type of associative learning is **operant conditioning** (or **trial-and-error learning**).

In operant conditioning, an animal learns to perform an act in order to receive a reward. This type of behavior was extensively studied by the psychologist, B. F. Skinner. He put a rat in an experimental cage and watched to see if it would randomly press different levers. Through trial and error, the animal figured out that one lever in particular would always produce food from a dispenser. Over time, the animal made an "association" between pressing the lever and getting food (the reward). Skinner even detected that some rats were so "conditioned" that they started to "hang out" near the lever. These same rats were also subjected to a negative form of operant conditioning, where touching the bar led to an electric shock. As you could imagine, rats quite quickly learned not to touch the bar that gave them the shock.

If the behavior is not reinforced, the conditioned response will be lost. This is called extinction. Here's one thing to remember:

> In **operant conditioning**, the animal's behavior determines whether it gets the reward or the punishment.

Habituation is another form of learning. It occurs when an animal learns not to respond to a stimulus. For example, if an animal encounters a stimulus over and over again without any consequences, the response to it will gradually lessen and may altogether disappear. One example is how a marine worm, *Nerels*, withdraws into its protective tube if a shadow passes over. With repeated exposure to the stimulus, however, the response decreases.

Insight

This is the highest form of learning and is exercised only by higher animals. **Insight** means the ability to figure out a behavior that generates a desired outcome. It is sometimes referred to as the "aha experience." As far as the animal world is concerned, human beings tend to be pretty good at using insight, or **reasoning**, to solve problems. However, we are not the only ones who reason. Chimpanzees, for example, have been known to use rudimentary tools, such as twigs and stones, to get their food.

To recap, there are four basic types of learning:

- **Imprinting**—occurs early in life and helps organisms recognize members of their own species.

- **Classical conditioning**—involves learning through association.

- **Operant conditioning**—occurs when a response is associated with new stimuli (also a form of associative learning).

- **Insight**—involves "reasoning" or problem solving.

INTERNAL CLOCKS: THE CIRCADIAN RHYTHM

There are other instinctual behaviors that occur in both animals and plants. One such behavior deals with time. Have you ever wondered how roosters always know when to start crowing? The first thought that comes to mind is that they've caught a glimpse of the sun. Yet many crow even before the sun has risen.

Roosters do have internal alarm clocks. Plants have them as well. These internal clocks, or cycles, are known as **circadian rhythms**.

If you've ever flown overseas, you know all about these. They're the basis of jet lag. Our bodies tell us it's one time while our watches tell us it's another. The sun may be up, but our body's internal clock is crying "Sleep!" This sense of time is purely instinctual: You don't need to know how to tell time in order to feel jet lag.

Circadian rhythms are yet another example of instinct. When an organism does something on a daily basis, we say it acts according to its circadian rhythm. But how do we know for certain that it's instinctual?

In a famous experiment, an American scientist took a bunch of plants and animals to the South Pole and put them on a turntable set to rotate at exactly the same speed as the earth but in the opposite direction of the earth's rotation. As a result, the organisms had absolutely no indication of day or night. Yet all of them continued to carry out their regular 24-hour cycles. This proved that the cycles have nothing to do with sunlight and everything to do with the internal clock.

Watch out though: Seasonal changes, like the loss of leaves by deciduous trees or the hibernation of mammals, are not examples of circadian rhythms. Circadian refers only to daily rhythms. Need a mnemonic? Just think how bad your jet lag would be after a trip around the world. In other words:

Circling the globe screws up your circadian rhythm.

HOW ANIMALS COMMUNICATE

Some animals use signals as a way of communicating with members of their species. These signals, which can be chemical, visual, electrical or tactile, are often used to influence mating and social behavior.

Chemical signals are one of the most common forms of communication among animals. Pheromones, for instance, are chemical signals between members of the *same* species that stimulate olfactory receptors and ultimately affect behavior. For example, when female insects give off their pheromones they attract males from great distances.

Visual signals also play an important role in the behavior observed among members of a species. For example, fireflies produce pulsed flashes that can be seen by other fireflies far away. The flashes are sexual displays that help male and female fireflies identify and locate each other in the dark.

Other animals use **electrical channels** to communicate. For example, some fish generate and receive weak electrical fields. Finally, **tactile signals** are found in animals that have mechanoreceptors in their skin to detect prey. For instance, cave-dwelling fishes use mechanoreceptors in their skin for communicating with other members and detecting prey.

Social Behavior

Many animals are highly social species and they interact with each other in complex ways. Social behaviors can help members of the species survive and reproduce more successfully. Several behavioral patterns for animal societies are summarized below:

- **Agonistic behavior** is aggressive behavior that occurs as a result of competition for food or other resources. Animals will show aggression toward other members that tend to use the same resources. A typical form of aggression is fighting between competitors.

- **Dominance hierarchies** (or pecking orders) occur when members in a group have established which members are the most dominant. The more dominant male will often become the leader of the group and usually have the best pickings of the food and females in the group. Once the dominance hierarchy is established, competition and tension within the group is reduced.

- **Territoriality** is a common behavior when food and nesting sites are in short supply. Usually the male of the species will establish and defend his territory (called a home range) within a group in order to protect important resources. This behavior is typically found among birds.

- **Altruistic behavior** is defined as unselfish behavior that benefits another organism in the group at the individual's expense because it advances the genes of the group. For example, when ground squirrels give warning calls to alert other squirrels of the presence of a predator, the calling squirrel puts itself at risk of being found by the predator.

ECOLOGY

The study of the interactions between living things and their environments is known as ecology. We've spent most of our time discussing individual organisms. However, in the "real world," organisms are in constant interaction with other organisms and the environment. The best way for us to understand the various levels of ecology is to progress from the big picture, the biosphere, down to the smallest ecological unit, the population.

Just as in taxonomy, there is a hierarchy within the "ecology world." Each of the following terms represents a different level of ecological interaction.

- **Biosphere**—The entire part of the earth where living things exist. This includes soil, water, light, and air. In comparison to the overall mass of the earth, the biosphere is relatively small. If you think of the earth as a basketball, the biosphere is equivalent to a coat of paint over its surface.

- **Ecosystem**—The interaction of living and nonliving things.

- **Community**—A group of populations interacting in the same area.

- **Population**—A group of individuals that belong to the same species and that are interbreeding.

BIOSPHERE

The biosphere can be divided into large regions called **biomes**. Biomes are massive areas that are classified mostly on the basis of their climates and plant life. Because of the different climates and terrains on the earth, the distribution of living organisms varies. For the AP Biology Exam, you're expected to know both the names of the different biomes and their characteristic **flora** (plant life) and **fauna** (animal life).

Here's a summary of the major biomes:

Major Biomes

Tundra
Regions—northernmost regions
Plant life—few, if any, trees; primarily grasses and wildflowers
Characteristics—contains permafrost (a layer of permanently frozen soil); has a short growing season
Animal life—includes lemmings, arctic foxes, snowy owls, caribou, and reindeer

Taiga
Region—northern forests
Plant life—wind-blown conifers (evergreens), stunted in growth, possess modified spikes for leaves
Characteristics—very cold, long winters
Animal life—includes caribou, wolves, moose, bear, rabbits, and lynx

Temperate Deciduous Forest
Regions—northeast and middle eastern U.S., western Europe
Plant life—deciduous trees that drop their leaves in winter
Characteristics—moderate precipitation; warm summers, cold winters
Animal life—includes deer, wolves, bear, small mammals, birds

Grasslands
Regions—American Midwest, Eurasia, Africa, South America
Plant life—grasses
Characteristics—hot summers, cold winters; unpredictable rainfall
Animal life—includes prairie dogs, bison, foxes, ferrets, grouse, snakes, and lizards

Deserts
Regions—western North America
Plant life—sparse, includes cacti, drought-resistant plants
Characteristics—arid, low rainfall; extreme diurnal temperature shifts
Animal life—includes jackrabbits (in North America), owls, kangaroo rats, lizards, snakes, tortoises

Tropical Rain Forests
Regions—South America, Colombia.
Plant life—high biomass; diverse types
Characteristics—high rainfall and temperatures; impoverished soil
Animal life—includes sloths, snakes, monkeys, birds, leopards, and insects

Remember that the biomes tend to be arranged along particular latitudes. For instance, if you hiked from Alaska to Kansas, you would pass through the following biomes: tundra, taiga, temperate deciduous forests, and grasslands.

ECOSYSTEM

Now we'll look at an even smaller group, the ecosystem. Ecosystems are self-contained regions that include both living and nonliving factors. For example, a lake, its surrounding forest, the atmosphere above it, and all the organisms that live in or feed off the lake would be considered an ecosystem. As you probably know, there is an exchange of materials between the components of an ecosystem. Take a look at the flow of carbon through a typical ecosystem:

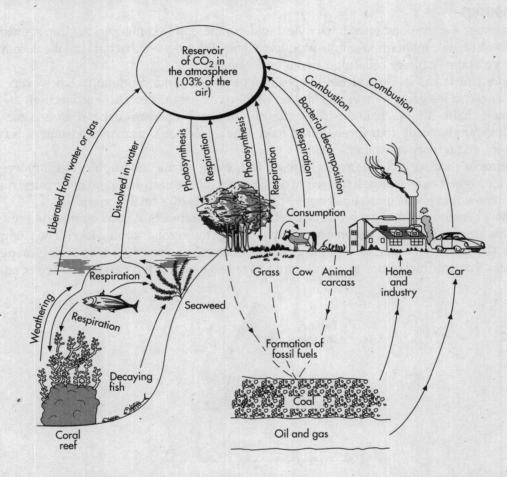

You'll notice how carbon is recycled throughout the ecosystem. In other words, energy flows through ecosystems. Other substances have similar cycles.

COMMUNITY

The next level we'll talk about is the community. A community refers to a group of interacting plants and animals that show some degree of interdependence. For instance, you, your dog, and the fleas on your dog are all members of the same community. All organisms within a community perform one of the following roles: **producers** (or autotrophs), **consumers** (or heterotrophs), or **decomposers**.

Producers

Producers, or autotrophs, have all of the raw building blocks to make their own food. From water and the gases that abound in the atmosphere, and with the aid of the sun's energy, autotrophs convert light energy to chemical energy. They accomplish this through **photosynthesis**.

Consumers

Consumers, or heterotrophs, are forced to find their energy sources in the outside world. Basically, heterotrophs digest the carbohydrates of their prey into carbon, hydrogen, and oxygen, and use these molecules to make organic substances.

The bottom line is: Heterotrophs, or "consumers," get their energy from the things they consume.

Decomposers

All organisms at some point must finally yield to decomposers. Decomposers are the organisms that break down organic matter into simple products. Generally, fungi and bacteria are the decomposers. They serve as the "garbage collectors" in our environment.

Each organism has its own **niche**—its position or function in a community. Since every species occupies a niche, it's going to have an effect on all the other organisms. These connections are shown in the **food chain**. A food chain describes the way different organisms depend on one another for food. There are basically four levels to the food chain: producers, primary consumers, secondary consumers, and tertiary consumers.

Autotrophs produce all of the available food. They make up the first **trophic** (feeding) level. They possess the highest biomass (the total weight of all the organisms in an area) and the greatest numbers. Did you know that plants make up about 99 percent of the earth's total biomass?

Primary consumers are organisms that directly feed on producers. A good example is a cow. These organisms are also known as **herbivores**. They make up the second trophic level. The energy flow, biomass, and numbers of members within an ecosystem can be represented in an **ecological pyramid**. Organisms that are "higher up" on the pyramid have less biomass and energy, and fewer numbers.

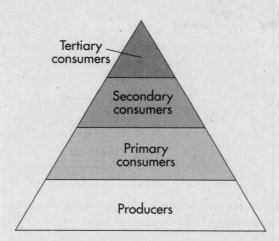

This simply means that primary consumers have less biomass and are fewer in number than producers. The next level consists of organisms that feed on primary consumers. They are the **secondary consumers**, and they make up the third trophic level. Above these are **tertiary consumers**.

So now you have it. We've got our four complete levels of the food chain.

- **Producers**—make their own food.
- **Primary consumers** (herbivores)—eat producers.

- **Secondary consumers** (carnivores and omnivores)—eat producers and primary consumers.

- **Tertiary consumers**—eat all of the above.

But what about decomposers? Where do they fit in on the food chain? They don't. They are not really considered part of the food chain. Decomposers are usually placed just below the food chain to show that they can decompose any organism.

In addition to the distinctions drawn above, remember:

> Toxins in an ecosystem are more concentrated and thus more dangerous for animals further up the pyramid.

This simply means that if a toxin is introduced into an ecosystem, the animals most likely to be affected are those at the top of the pyramid. This occurs because of the increasing concentration of such toxins. The classic example of this phenomenon is DDT, an insecticide initially used to kill mosquitoes. Although the large-scale spraying of DDT resulted in a decrease in the mosquito population, it also wound up killing off ospreys.

Ospreys are aquatic birds whose diet consists primarily of fish. The fish that ospreys consumed had, in turn, been feeding on contaminated insects. Since fish eat thousands of insects, and ospreys hundreds of fish, the toxins grew increasingly concentrated. Though the insecticide seemed harmless enough, it resulted in the near-extinction of certain osprey populations. What no one knew was that in sufficient concentrations, DDT weakened the eggshells of ospreys. Consequently, eggs broke before they could hatch, killing the unborn ospreys.

Such environmental tragedies still occur. All ecosystems, small or great, are intricately woven, and any change in one level invariably results in major changes at all the other levels.

SYMBIOTIC RELATIONSHIPS

Many organisms that coexist exhibit some type of symbiotic relationship. These include remoras, or "sucker fish," which attach themselves to the backs of sharks, and lichen, the fuzzy, mold-like stuff that grows on rocks and many other examples. Lichen appears to be one organism, when in fact it is two organisms—a fungus and an alga or photosynthetic bacterium—living in a complex symbiotic relationship.

Overall, there are three basic types of symbiotic relationships:

- **Mutualism**—in which both organisms win (for example, the lichen).

- **Commensalism**—in which one organism lives off another with no harm to the "host" organism (for example, the remora).

- **Parasitism**—in which the organism actually harms its host.

Regardless of which one we refer to, all three are examples of "symbiotic" relationships. Make sure you're clear on the differences among the three.

We've found a great way to sum up the different types of symbiosis:

Relationship	Organism I	Organism II
Commensalism	😐	🙂
Parasitism	🙁	🙂
Mutualism	🙂	🙂

POPULATION ECOLOGY

Population ecology is the study of how populations change. Whether these changes are long-term or short-term, predictable or unpredictable, we're talking about the growth and distribution patterns of a population.

When studying a population, you need to examine four things: the **size** (the total number of individuals), the density (the number of individuals per area), the **distribution patterns** (how individuals in a population are spread out), and the **age structure**.

One way to understand the growth pattern and make predictions about the population growth of a country is by examining **age structure histograms**. For example, in underdeveloped countries, where the population increase is high, the base of the histogram is very wide compared to countries that show moderate growth.

Another way to study the changes in a population is by looking at **survivorship curves**. These curves are graphs of the numbers of individuals surviving to different ages, indicating the probability of any individual living to a given age.

For example, the graph below shows there is a high death rate among the young of oysters but those that survive do well. On the other hand, there is a low death rate among the young of humans, but, after age 60, the death rate is high.

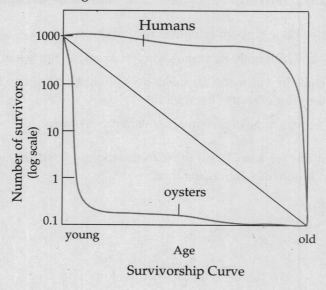

Survivorship Curve

The growth of a population can be represented as crude birth – crude death/size of the population:

$$r = \text{births} - \text{deaths}/N$$

(r is the reproductive rate, and N is the population size)

Each population has a **carrying capacity**—the maximum number of individuals of a species that a habitat can support. Most populations, however, don't reach their carrying capacity because they're exposed to limiting factors.

One important factor is **population density**. The factors that limit a population are either density-independent or density-dependent. **Density-independent factors** are factors that affect the population regardless of the density of the population. Some examples are severe storms and extreme climates. On the other hand, **density-dependent factors** are those whose effects depend on population density. Resource depletion, competition, and predation are all examples of density-dependent factors. In fact, these effects become even more intense as the population density increases.

Exponential growth

The growth rates of populations also vary greatly. There are two types of growth: exponential growth and logistic growth. **Exponential growth** occurs when a population is in an ideal environment. Growth is unrestricted because there are lots of resources, space, and no disease or predation. Here's an example of exponential growth. Notice that the curve arches sharply upward—the exponential increase.

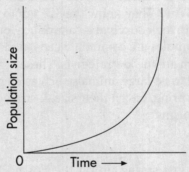

Exponential growth occurs very quickly, resulting in a **J-shaped curve**. A good example of exponential growth is the initial growth of bacteria in a culture. There's plenty of room and food, so they multiply rapidly—every 20 minutes.

However, as the bacterial population increases, the individual bacteria begin to compete with each other for resources. The population reaches its carrying capacity and the curve levels off.

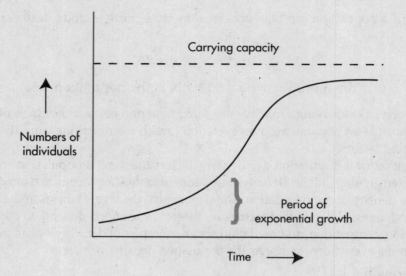

The population becomes restricted in size because of limited resources. This is referred to as **logistic growth**. Notice that the growth forms an S-shaped curve. These growth patterns are associated with two kinds of life-history strategies: **r-selected species** and **k-selected species**.

We've already mentioned that organisms that grow exponentially approach the carrying capacity. These organisms tend to thrive in areas that are barren or uninhabited. Once they colonize an area they reproduce as quickly as possible. Why? They know they've got to multiply before competitors arrive on the scene! The best way to ensure their survival is to produce lots of offspring. These organisms are known as **r-strategists**. Typical examples are common weeds, dandelions, and bacteria.

At the other end of the spectrum are the **k-strategists**. These organisms are best suited for survival in stable environments. They tend to be large animals such as elephants, with long lifespans. Unlike r-strategists, they produce few offspring. Given their size, k-strategists usually don't have to contend with competition from other organisms.

ECOLOGICAL SUCCESSION

Communities of organisms don't just spring up on their own; they develop gradually over time. **Ecological succession** refers to the predictable procession of plant communities over a relatively short period of time (decades or centuries).

Centuries may not seem like a short time to us, but if you consider the enormous stretches of time over which evolution occurs, hundreds of thousands or even millions of years, you'll see that it is pretty short.

The process of ecological succession where no previous organisms have existed is called **primary succession**.

The pioneers

How does a new habitat full of bare rocks eventually turn into a forest? The first stage of the job usually falls to a community of lichens. Lichens are hardy organisms. They can invade an area, land on bare rocks and erode the rock surface, and over time turn it into soil. Lichens are considered **pioneer organisms**.

Once lichens have made an area more habitable, they've set the stage for other organisms to settle in. Communities establish themselves in an orderly fashion. Lichens are replaced by mosses and ferns, which in turn are replaced by tough grasses, then low shrubs, then deciduous trees, and finally, the evergreen trees. Why are lichens replaced? Because they can't compete with the new plants for sunlight and minerals.

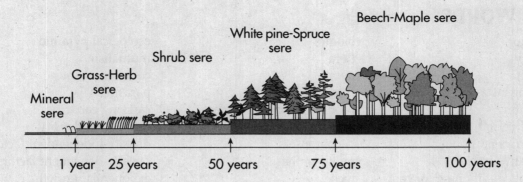

The entire sequence is called a **sere**. The final community is called the **climax community**. The climax community is the most stable. In our example, the evergreens are part of the climax community.

Now what happens when a forest is devastated by fire? The same principles apply but the events occur much more rapidly. The only exception is that the first invaders are usually not lichens but grasses, shrubs, saplings, and weeds. When a new community develops where another community has been destroyed or disrupted, this event is called secondary succession.

HUMAN IMPACT ON THE ENVIRONMENT

Unfortunately, humans have disturbed the existing ecological balance and the results are far-reaching. Soils have been eroded and various forms of pollution have increased. The potential consequences on the environment are summarized below:

- **Greenhouse effect**—The increasing atmospheric concentrations of carbon dioxide through the burning of fossil fuels and forests have led to an increase in the warming of the earth. Higher temperatures may cause the polar ice caps to melt and flooding to occur.

- **Ozone depletion**—Pollution has also led to the depletion of the atmospheric ozone layer by such chemicals as clorofluorocarbons, which are used in aerosol cans. Ozone protects the earth's surface from excessive ultraviolet radiation. Its loss could have major genetic effects and could increase the incidence of cancer.

- **Acid rain**—Pollutants, such as sulfur dioxide and nitrogen, from the burning of fossil fuels has lowered the pH of rain. Acid rain damages crops and kills animals.

- **Desertification**—When land is overgrazed by animals it turns grasslands into deserts.

- **Deforestation** –When forests are cleared (especially by the slash and burn method) erosion, floods, and changes in weather patterns can occur.

- **Pollution**—Some pollutants, such as DDT, were overused at one time, damaging many plants and animals. Since DDT resists chemical breakdown, today it is found in the tissues of nearly every living organism. DDT has threatened some animals with extinction.

- **Reduction in species diversity**—As different habitats have been destroyed, many plants and animals have become extinct. Some of these plants could have provided us with medicines and products that would have been beneficial.

KEY WORDS

behavior
ecology
instinct
learning
imprinting
critical period
classical conditioning
associative learning
operant conditioning (or trial-
    and-error learning)
insight
reasoning
circadian rhythms
ecology
biosphere
ecosystem
community
population

biomes
flora
fauna
community
producers (or autotrophs)
consumers (or heterotrophs)
decomposers
photosynthesis
niche
food chain
trophic
primary consumers
secondary consumers
herbivores
acid rain
deforestation
ozone depletion
desertification

ecological pyramid
mutualism
commensalism
parasitism
carrying capacity
population density
density-independent factors
density-dependent factors
exponential growth
logistic growth
r-strategists
k-strategists
ecological succession
pioneer organisms
sere
climax community
greenhouse effect
pollution

QUIZ

<u>Directions:</u> Each of the questions or incomplete statements below is followed by five suggested answers or completions. Select the one that is best in each case. Answers can be found on page 246.

1. When a seed is set in the ground, roots will grow downward while stems will grow upward. This plant behavior is an example of

   (A) hydrotropism.
   (B) chemotropism.
   (C) gravitropism.
   (D) thigmotropism.
   (E) circadian rhythm.

2. Which of the following systems most directly regulate behavioral responses in animals?

   (A) Excretory and immune systems
   (B) Digestive and endocrine systems
   (C) Nervous and skeletal systems
   (D) Reproductive and nervous systems
   (E) Endocrine and nervous systems

Questions 3–5 refer to the following food chain.

Birds → snakes → mice → insects → plants

3. Which of the following organisms in this food chain can transform light energy to chemical energy?

    (A) Birds
    (B) Insects
    (C) Plants
    (D) Snakes
    (E) Mice

4. Which animal in the food chain has the smallest biomass?

    (A) Birds
    (B) Insects
    (C) Plants
    (D) Snakes
    (E) Mice

5. Which organism in the food chain is a secondary consumer?

    (A) Birds
    (B) Insects
    (C) Plants
    (D) Snakes
    (E) Mice

6. The portion of the earth that is inhabited by life is known as the

    (A) ecosystem.
    (B) biosphere.
    (C) biome.
    (D) population.
    (E) community.

7. Which of the following organisms serve as decomposers in the ecosystem?

    (A) Bacteria and viruses
    (B) Fungi and bacteria
    (C) Viruses and protists
    (D) Fungi and viruses
    (E) Bacteria and plants

Directions: Each group of questions consists of five lettered headings followed by a list of numbered phrases or sentences. For each numbered phrase or sentence, select the one heading that is most closely related to it and fill in the corresponding oval on the answer sheet. Each heading may be used once, more than once, or not at all in each group.

Questions 8–10

(A) Instinct
(B) Operant learning
(C) Imprinting
(D) Tropism
(E) Insight

8. A stickleback fish will not attack an intruder that lacks a red belly

9. A bluejay avoids monarch butterflies after experiencing their distasteful poisoning

10. A chimpanzee uses several boxes on the floor to reach bananas hung from the ceiling

Questions 11–14

(A) Taiga
(B) Tundra
(C) Desert
(D) Tropical rain forest
(E) Temperate deciduous forest

11. The biome at the highest altitudes

12. The biome characterized by harsh winters, short summers, and evergreen trees

13. The biome with plants having low surface-to-volume ratios

14. The biome with the greatest species diversity

14

The Free-Response Questions

THE ESSAYS

We briefly discussed the essay portion of the test—better known as "the babble"—in Chapter 1. We saw that Part II of the AP Biology Exam consists of four essays, one or more of which are divided into several parts. One question will come from area 1, Molecules and Cells, another question will come from area 2, Heredity and Evolution, and two questions will come from area 3, Organisms and Populations. You'll have 10 minutes to read and 90 minutes to complete these essays. Now, let's look at the essays in greater detail.

At first glance, most students panic when they look at the essay questions on the test. They wonder how they are going to write four well-thought-out essays, complete with figures and graphs, in 90 minutes! They believe this must be the toughest part of the test. Well, they're wrong. Believe it or not, this is the part of the test on which you can actually shine. Why? Because you're no longer limited to multiple-choice questions where there's only one right answer. This time, you can dazzle the test reviewers with your vast knowledge of biology (especially after having read this book). We've come up with six tips that will help you ace the essay portion of the test.

TIP 1: READ THE QUESTIONS CAREFULLY

Starting in June 2004, ETS is giving you 10 minutes to read the questions and organize your thoughts before you begin writing. If you use these 10 minutes wisely, you can breeze your way through the essays. The first thing you should do is take less than a minute to skim all of the questions and put them into your own personal order of difficulty from easiest to toughest. Once you've decided the order in which you will answer the questions (easiest first, hardest last), you can begin to formulate your responses. Your first step should be a more detailed assessment of each question.

The most important advice we can give you is to read each question at least twice. As you read the question, focus on key words, especially "direction words." Almost every essay question begins with a direction word. Some examples of direction words are *discuss*, *define*, *explain*, *describe*, *compare*, and *contrast*. If a question asks you to discuss a particular topic, you should give a viewpoint and support it with examples. If a question asks you to compare two things, you should discuss how the two things are similar. On the other hand, if the question asks you to contrast two things, you need to show how these things are different.

Many students lose points on their essays because they either misread the question or fail to do what's asked of them.

TIP 2: BRAINSTORM

Your next objective during the 10-minute preview should be to organize your thoughts.

Once you've read the question, you need to brainstorm. Jot down as many key terms and concepts—the "hot buttons" we mentioned in chapter 1—as you can. Don't forget, the test reviewers assign points on the basis of these key concepts: For each one that you mention and/or explain, you get a point.

How many do you need? You won't need all of them, that's for sure. However, you will need enough to get you the maximum number of points for that question. Most of these you can pull directly from your reading in this book. Remember the "Key Words" lists in each chapter of this book? They are essential in helping you to prepare for this part of the test. Use your word associations to generate thorough lists. Once you've come up with as many different hot buttons as possible, you're ready to leap into your outline.

Don't spend more than about 2 minutes per question brainstorming. Once your 10-minute reading period is up you should be ready to start writing your essays.

TIP 3: OUTLINE YOUR ESSAY

Have you ever written yourself into a corner? You're halfway through your essay when you suddenly realize that you have no idea whatsoever where you're going with your train of thought. To avoid this (and the panic that accompanies it), take a few minutes to draft an outline.

Your outline should incorporate as many of the hot buttons as you need in order to maximize your score. In other words, if they ask for two examples, choose only those two with which you are most comfortable. In your outline, make notes about the crucial points to mention with regard to each topic or key word. Once your outline is complete, you're ready to move on to writing the actual essay.

All of this preparation may seem time-consuming. However, it should take no more than four or five minutes per essay. What's more, it will greatly simplify the whole essay-writing process. So while you lose a little time at the outset, you'll more than make up for it when it comes time to actually write your answer.

> Remember: Unless you're aiming to score a perfect 5, there's no reason to do all the essays. If you're uncertain about how many you need to reach your score, look back at the pacing chart in chapter 1.

TIP 4: DEVELOP YOUR IDEAS IN EACH ESSAY

Now you can use your outline to write your essay. Most students can come up with key terms or phrases that concern a particular biology topic. What separates a high-scoring student from a low-scoring student is *how* the student develops his or her thoughts on each essay. Besides giving the hot buttons, you'll need to elaborate on your thoughts and ideas. For example, don't just throw out a list of terms that pertain to meiosis and mitosis (like synapsis, crossing-over, and gametes). Go one step further. Make sure you mention the *significance* of meiosis (i.e., it produces genetic variability). This extra piece of information will earn you an extra point.

Generally, you'll need to write about two to four paragraphs, depending on the number of parts contained in each essay question. In addition, be sure to give the appropriate number of examples for each essay. If the question asks for three examples, give only three examples. If you present more than is required, the test reviewers won't even read them or count them toward your score. The bottom line is this:

> Stick to the question.

TIP 5: ANSWER EACH PART OF THE ESSAY QUESTION SEPARATELY

The maximum number of points you can get on each essay is 10. Consequently, if a question has four parts, assume that each section is worth about 2.5 points. The more parts there are to an essay question, the more important it is to pace yourself. On each essay, you're better off writing a little bit for each part than you are spending all your time on any one part of a question. Why? Even if you were to write the perfect answer to one part of a question, there's a limit to the number of points the test reviewers can assign to that part. Moreover, by writing a separate paragraph for each section, you make the test grader's job that much easier. When test readers have an easy time reading your essay, they're more likely to award points: It comes across as clearer and more organized.

Finally, don't spend too much time writing a fancy introduction. You won't get brownie points for beautifully written openings like, "It was the best of experiments, it was the worst of experiments." Just leap right into the essay. And don't worry too much about grammar or spelling errors. Your grammar can hurt only if it's so bad that it seriously impairs your ability to communicate.

TIP 6: INCORPORATE ELEMENTS OF AN EXPERIMENTAL DESIGN

Since one of the essay questions will be experimentally based, you'll need to know how to design an experiment. Most of these questions require that you present an appropriately labeled diagram or graph. Otherwise, you'll only get partial credit for your work.

There are two things you must remember when designing experiments on the AP Biology Exam: (1) Always label your figures, and (2) include controls in all experiments. Let's take a closer look at these two points.

Know how to label diagrams and figures

Let's briefly discuss the important elements in setting up a graph. The favorite type of graph on the AP Biology Exam is the *coordinate graph*. The coordinate graph has a horizontal axis (*x-axis*) and a vertical axis (*y-axis*).

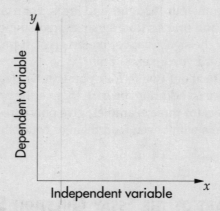

The *x*-axis usually contains the *independent variable*—the thing that's being manipulated or changed. The *y*-axis contains the *dependent variable*—the thing that's affected when the independent variable is changed.

Now let's look at what happens when you put some points on the graph. Every point on the graph represents both an independent variable and a dependent variable.

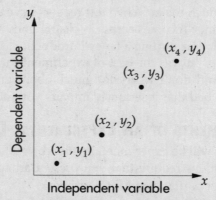

Once you draw both axes and label the axes as x and y, you can plot the points on the graph. In Chapter 1, we saw an essay question that included a part about enzymes and the influence of various factors on enzyme activity. Let's look at the question again.

1. Enzymes are biological catalysts.

   a. Relate the chemical structure of an enzyme to its catalytic activity and specificity.

   b. Design an experiment that investigates the influence of temperature, substrate concentration, or pH on the activity of an enzyme.

   c. Describe what information concerning enzyme structure could be inferred from the experiment you have designed.

For now, let's discuss only part *b*, which asks us to design an experiment. Let's set up a graph that shows the results of an experiment examining the relationship between pH and enzyme activity. Notice that we've chosen only one factor here, pH. We could have chosen any of the three. Why did we choose pH and not temperature or substrate concentration? Well, perhaps it's the one we know the most about.

What is the independent variable? It is pH. In other words, pH is being manipulated in the experiment. We'll therefore label the *x*-axis with pH values from 0 through 14.

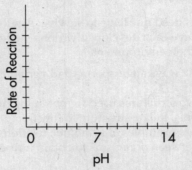

What is the dependent variable? It's the enzyme activity—the thing that's affected by pH. Let's label the y-axis "Rate of Reaction." Now we're ready to plot the values on the graph. Based on our knowledge of enzymes, we know that for most enzymes the functional range of pH is narrow, with optimal performance occurring at or around a pH of 7.

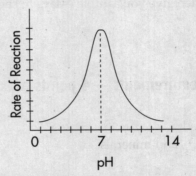

Now you should interpret your graph. If the pH level decreases from a neutral pH of 7, the reaction rate of the enzyme will decrease. If the pH level increases, the rate of reaction will also decrease. Don't forget to include a simple explanation of your graph.

Include controls in your experiments

Almost every experiment will have at least one variable that remains constant throughout the study. This is called the *control*. A control is simply a standard of comparison. What does a control do? It enables the biologist to verify that the outcome of the study is due to changes in the independent variable and nothing else.

Let's say the principal of your school thinks that students who eat breakfast do better on the AP Biology Exam than those who don't eat breakfast. He gives a group of 10 students from your class free breakfast every day for a year. When the school year is over, he administers the AP Biology Exam and they all score brilliantly! Did they do well because they ate breakfast every day? We don't know for sure. Maybe the principal hand picked the smartest kids in the class to participate in the study.

In this case, the best way to be sure that eating breakfast made a difference is to have a control group. In other words, he would need to pick students in the class who *never* eat breakfast and follow them for a year. At the end of that year, he could send them in to take the AP Biology Exam. If they do just as well as the group that ate breakfast, then we can probably conclude that eating breakfast wasn't the only factor leading to higher AP scores. The group of students that didn't eat breakfast is called the control group because those students were not "exposed" to the variable of interest—in this case, breakfast.

Now that we've covered the important points on how to tackle the essay section, let's see if you can write a good essay using the techniques we've just discussed. Take 22 minutes to write a response to the sample essay.

1. All organisms need nutrients to survive. Angiosperms and vertebrates have each developed various methods to obtain nutrients from their environment.

    a. Discuss the ways angiosperms and vertebrates procure their nutrients.
    b. Discuss two structures used for obtaining nutrients among angiosperms. Relate structure to function.
    c. Discuss two examples of symbiotic relationships that have evolved between organisms to obtain nutrients.

GRADING CHECKLIST

To help you grade this sample essay, we've put together a checklist that you can use to calculate the number of points that should be assigned to each part of this question. We'll first explain the important points on the checklist, and then give you sample essays to show you how test reviewers would evaluate them.

ESSAY CHECKLIST

Part a: Types of nutritional requirements — 4 points maximum

Angiosperms:
1 point—Autotrophs—if defined
1 point—They require H_2O, CO_2, and minerals.

Vertebrates:
1 point—Heterotrophs—if defined
1 point—They require organic compounds (sugars, proteins, fats), H_2O, vitamins, and minerals.

Part b: Structures in angiosperms — 4 points maximum

First two structures only

(1 point each)	(1 point each)
Structure	**Function described**
Cuticles	Prevent desiccation
Stomata	Pores in leaf that regulate gas exchange
Lenticels	Pores in stems that regulate gas exchange
Palisade mesophyll	Loosely packed cells in a leaf; contains chloroplasts
Roots	Structures that absorb water and minerals
Root hairs	Extensions of the root that increase the surface area for water absorption

(1 point)

Carnivorous plants (such as Venus flytrap) have modified leaves to trap insects.

Part c: Symbiotic relationships — 4 points maximum

First 2 relationships only

(1 point each)	(1 point each)
<u>Relationship</u>	<u>Benefits to Each</u>
Nitrogen-fixing bacteria	Bacteria get food/Plants get useable nitrogen
Mycorrhizae	Fungi get food/Plants get water and minerals
Intestinal bacteria	Humans get Vitamin K/Bacteria have a place to live
Ruminants flora	Bacteria digest the cellulose for cows/Bacteria have a place to live
Wood-eating termites	Protists digest wood/Termites obtain the wood

EXPLANATION OF CHECKLIST

For Part a, you'll receive points if you demonstrate a clear understanding of the terms autotroph *and* heterotroph. You must define the terms and describe the necessary nutrients for both angiosperms and vertebrates. For angiosperms, you have to mention the raw materials for photosynthesis (CO_2 + H_2O) and minerals. For vertebrates, you have to mention organic compounds and additional vitamins or minerals.

For Part b, you have to relate structure to function. Which structures are involved in obtaining nutrients? You're asked to discuss *two* structures in angiosperms. One point is given for the structure and another point is given if you show a clear understanding of the function of the structure.

For Part c, you are asked to give *two* examples of symbiotic relationships. Since they ask for only two examples, you earn points only for the first two examples. An additional point is also given if you describe the benefits to each organism.

Notice that you don't have to know everything about biology to get a high score on the essay section. However, you do have to present the proper information with a certain level of detail. Additionally, you can earn a maximum of 4 points for each part. However, you can't earn more than 10 points total.

Before you use this checklist to assign points to your essay, let's evaluate two sample essays.

SAMPLE ESSAYS

Let's read an essay written by Joe Bloggs, which was given 4 points. (All of Joe's grammar and spelling mistakes were retained.)

JOE BLOGGS'S ESSAY

Angiosperms are autotrophs which make their own food. They have chloroplasts within the leaves which captures sunlight. Angiosperms also get nutrients through their roots which absorb water and bring it up the leaf. Food, water and minerals are then transport in the plant by xylem and phloem. Vertebrates can not make their own food so they must get their nutrients from other organisms. Some vertebrates only eat plants while other vertebrates eat both plants and animals.

Two structures that angiosperms use to obtain nutrients from the environment are leaves and roots. Plants have leaves which are shaped to absorb the maximum amount of sunlight. Plants also have root hairs which increase the surface area so that more water and minerals can be taken up the leaf.

An example of a symbiotic relationship is the bacteria found in the root nodules of plants. Plants need nitrogen to make plant proteins but they can't use the nitrogen from the atmosphere. Plants therefore work with nitrogen-fixing bacteria which convert nitrogen to nitrates. Both organisms benefit from this relationship.

Explanation

Joe Bloggs would probably get a 4 for this essay. For part a, he clearly defined the term autotroph and received 1 point. However, he did not list the starting materials (CO_2 + H_2O) to make food in plants. He also failed to define heterotrophs. Consequently, he didn't receive any points for his explanation of heterotrophs. Notice that for this paragraph, he wasn't penalized for any grammar mistakes.

For part b, Joe received 2 points for discussing root hairs and their function. Unfortunately, his explanation of leaves was not detailed enough, so he did not receive any additional points.

For part c, Joe received 1 point for his example of a symbiotic relationship—nitrogen-fixing bacteria and plants. However, he did not specifically describe the benefits to *each* organism. Although he mentioned the benefits to plants, he neglected to discuss the benefits to bacteria. Had he completed his thought, he would have earned another point. Overall, although his essay was a little choppy, Joe was not penalized for it.

Now let's read an essay written by Josephine Bloggs, which was given 9 points.

JOSEPHINE BLOGGS'S ESSAY

Organisms have developed a variety of ways to obtain nutrients. Angiosperms, which are flowering plants, are autotrophs. This means that they are capable of making their own food via photosynthesis. The reaction for photosynthesis is:

$$6CO_2 + 12\,H_2O \xrightarrow{\text{sunlight}} C_6H_{12}O_6 + 6O_2 + 6H_2O + ATP$$

Plants use carbon dioxide and water to produce glucose and ATP. Angiosperms also use their roots to absorb water and essential minerals from the soil.

Vertebrates, however, are heterotrophs. They are unable to make their own food. They must therefore rely on other organisms to obtain energy. They require organic compounds, such as proteins and carbohydrates, and break them down to simpler molecules that can be absorbed into the body. Vertebrates also require other essential substances such as vitamins, which serve as coenzymes in crucial biological reactions.

Two structural adaptations used by angiosperms to obtain nutrients are root hairs and stomates. Plants contain numerous root hairs which are tiny projections from a plant's root that increase the surface area for the uptake of water and minerals. Plants also have stomates which regulate the intake of CO_2. The carbon dioxide that enters the stomates can then be used to make glucose.

There are many examples of symbiotic relationships. A common example is nitrogen fixation. Certain plants, called legumes, require nitrogen to synthesize plant amino acids. These plants have nitrogen-fixing bacteria which live in their root nodules and convert nitrogen to nitrates. The plant uses the nitrates to make plant proteins. The bacteria, in turn, gets energy from the carbohydrates of the plant.

Another example of a symbiotic relationship is bacteria in the stomach of cows. Cows are herbivores that can not digest cellulose. They rely on bacteria in their stomach to digest some of the cellulose found in their stomach. Cows, therefore, get some of their energy from fermenting bacteria. A third example of symbiosis is the bacteria found in vertebrates, such as humans. Humans have bacteria that reside in their large intestine. These microorganisms break down undigested food particles and produce Vitamin K. The bacteria in turn have a place to live.

Explanation

This essay received 9 points. Josephine clearly demonstrated a mastery of the material. In part a, she defined and discussed the terms autotroph and heterotroph. She also mentioned the important raw materials that are involved in each case. Notice that she presented and explained the equation for photosynthesis. She received 4 points—the maximum—for this section.

In part b, Josephine presented two structures and their function. Her coverage of the material was fairly thorough. She talked about two important structures: root hairs and stomates. She received 2 points, one for mentioning each structure, and 1 point for her discussion of root hairs. She didn't mention the structure of stomates—that they're pores in the epidermis—so she did not receive a point for the function of stomates.

For part c, Josephine presented three types of symbiotic relationships. The reader reviewed only the first two examples to assign points. Josephine received 2 points for her discussion of nitrogen fixation but only 1 point for her explanation of bacteria in the cow's stomach. In this case, she forgot to mention the benefits to bacteria.

HOW TO USE OUR PRACTICE ESSAYS

The most important advice for this section of the test is *PRACTICE, PRACTICE, PRACTICE!* No matter how well you think you know the material, it's important to practice formulating your thoughts on paper. Before you take our practice tests in the back of the book, make sure you review our techniques for the essay portion. Then use our checklist—which is put together just like ETS's—to assign points.

To give you as much practice as possible, we've compiled a few sample AP biology questions. In fact, most of these questions refer to themes in biology that crop up time and time again. More often than not, you'll see one of these questions (or something similar) on your exam.

SAMPLE ESSAY QUESTIONS

1. Describe how the modern theory of evolution is supported by evidence from the following areas:
   a. Comparative anatomy
   b. Biogeography
   c. Embryology

2. Describe the structure of a generalized eukaryotic plant cell. Contrast the structure of a nonphotosynthetic prokaryotic cell to a generalized eukaryotic plant cell.

3. Discuss the following responses in plants and give one example of each.
   a. Photoperiodism
   b. Phototropism
   c. Geotropism

4. Describe the chemical nature of genes. Discuss the replicative process of DNA in eukaryotic organisms. What are the various types of gene mutations that can occur during replication?

5. Select two of the following pairs of hormones and discuss the concept of negative feedback.

    a. Thyroid-stimulating hormone (TSH) and thyroxine
    b. Parathyroid hormone and calcitonin
    c. Cortisol and ACTH

6. Describe the process of photosynthesis. What are the major plant pigments involved in photosynthesis? Design an experiment to measure the rate of photosynthesis.

7. Discuss the mechanism by which a muscle cell contracts. Relate structure to function.

The biology you need to answer all of these essay questions is in this book. After you have completed your essay(s), go back to the chapter(s) that cover these topics and check to see if you have presented all the relevant points. Use our Essay Checklist as a guideline to give yourself an approximate score. Good luck!

15

Laboratory

All AP biology courses have a laboratory component that gives students hands-on experience regarding some of the biology topics covered in class. Through these laboratory exercises you can learn the scientific method, lab techniques, and problem-solving skills. The folks at the College Board require that you complete 12 laboratory exercises during the school year. These labs are supposed to draw from some of the most important topics in biology.

How do these labs relate to the AP Biology Exam? Approximately 10 percent of the questions on the multiple-choice section and one essay question will refer to certain aspects of these 12 lab exercises. ETS incorporates questions concerning these labs to determine two things: (1) how well you understand the key concepts in biology, and (2) how well you think analytically. ETS wants to find out if you can design experiments, manipulate data, and draw conclusions from experiments.

There are basically two ways to approach a review of the laboratory exercises for this exam. We could review every single experiment, including all the procedures and results. This is what ETS expects you to do. Not only would this take up more than half of this book, it wouldn't necessarily improve your score.

We at The Princeton Review have a better approach. Believe it or not, many schools do not cover all of the laboratory experiments. Why? Some schools don't have the proper facilities to conduct the more complicated experiments. Other schools have opted to use their own exercises, which illustrate the same concepts. Given this, we'd rather not overwhelm you with all the boring details about each and every experiment. Instead, we're going to tell you the *objective* of each lab. That way, if you conducted a different lab (or if you've never even seen the lab), it doesn't matter. We'll bring to light the important concepts and results related to each experiment.

In addition, some of these labs cover techniques or terms that are usually taught during labs. So pay particular attention to these exercises. How should you use this chapter to review for the AP Biology Exam? Read each summary and focus on the *concepts* stressed in each experiment. Now, let's get cracking.

LAB 1: DIFFUSION AND OSMOSIS

This lab investigated the process of diffusion and osmosis in a semi-permeable membrane as well as the effect of solute concentration on water potential in plants.

What are the general concepts you really need to know?

- Fortunately, this lab covered the same concepts about diffusion and osmosis that are discussed in this book. Just remember that osmosis is the movement of water across a semi-permeable membrane from a region of high water concentration to one of low water concentration (from a region of low solute concentration to a region of high solute concentration).

- Be familiar with the concept of *water potential*. Water potential is simply the free energy of water. It is a measure of the tendency of water to diffuse across a membrane. Water moves across a selectively permeable membrane from an area of higher water potential to an area of lower water potential.

- Be familiar with the effects of water gain in animal and plant cells. In animals, the direction of osmosis depends on the concentration of solutes both inside and outside of the cell. In plants, osmosis is also influenced by *turgor pressure*—the pressure that develops as water presses against a plant's cell wall.

LAB 2: ENZYME CATALYSIS

In this lab, the rate of conversion of hydrogen peroxide (H_2O_2) to water and oxygen gas was measured as the enzyme catalase was added to the experiment.

$$H_2O_2 \xrightarrow{\text{catalase}} H_2O + \frac{1}{2} O_2 \text{ (gas)}$$

What are the general concepts you *really* need to know?

- This lab tested your basic understanding of enzymes. Enzymes are catalysts that accelerate specific reactions without altering the reaction itself.

- Don't forget that reaction rates can be influenced by four things: pH, temperature, enzyme concentration, and substrate concentration.

- Remember that a reaction can stop if the enzyme is denatured, that is, if it loses its shape.

- If you're asked to design an experiment to measure the effect of these four variables on enzyme activity, keep all the conditions constant except for the variable of interest. For example, to measure the effects of pH in an experiment, maintain the temperature, enzyme concentration, and substrate concentration as you change the pH.

LAB 3: MITOSIS AND MEIOSIS

This lab highlighted the differences between mitosis and meiosis. There were two parts to this lab.

EXERCISE 3A

In lab 3A, slides of onion root tips were prepared to study plant mitosis while slides of a white fish blastula were prepared to study animal mitosis.

What are the general concepts you *really* need to know?

Be familiar with the concept of mitosis in both plants and animals. In plants, primary growth is limited to dividing cells in the tips of roots and stems. In animals, however, mitosis occurs in all growing cells, especially those in dividing embryos.

EXERCISE 3B

In lab 3B, the sexual life cycle of the fungus *Sordaria fimicola* was examined. Sexual reproduction in this fungus involves the fusion of two nuclei—a plus strain and a minus strain—to form a diploid zygote. The zygote immediately undergoes meiosis, and haploid cells called *ascospores* are produced. These ascospores are found in tiny, sac-like structures called asci.

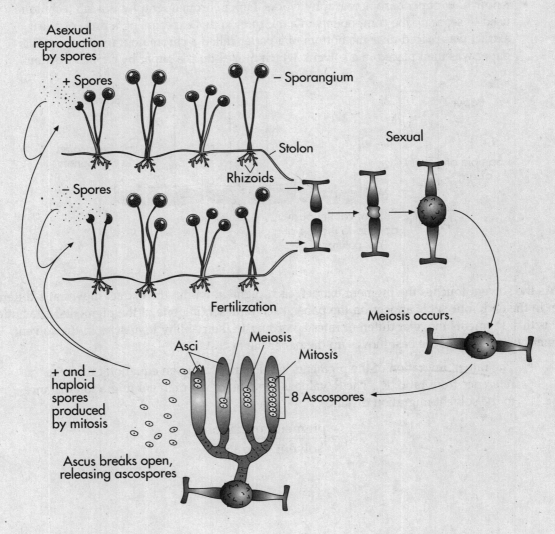

What are the general concepts you really need to know?

- Know the concept of meiosis. Meiosis involves crossing-over of chromosomes to create genetic variability. Crossing-over in fungi results in a particular arrangement of these ascospores within asci. If crossing-over occurs, you'll see *different* genetic combinations in the offspring compared to the parent strain. If, however, crossing-over does not occur, each ascospore will be *identical* to the parent strain.

LAB 4: PLANT PIGMENTS AND PHOTOSYNTHESIS

This lab has two parts.

EXERCISE 4A

In lab 4A, four plant pigments—chlorophyll *a*, chlorophyll *b*, xanthophylls, and carotenoids—were separated using paper chromatography.

What are the general concepts you *really* need to know?

- Know how paper chromatography works. Paper chromatography is a lab technique used to separate the components in a mixture. In this experiment, a drop of a leaf extract was placed near the bottom of a paper called a chromatography paper. This paper was then placed in a solvent, which moves up the paper by capillary action.

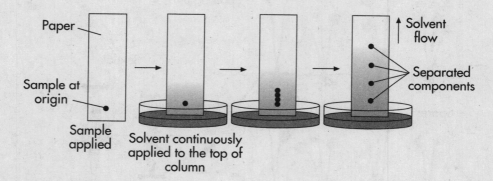

As the solvent touches the pigment extract, each pigment within the extract moves at a different rate. In the end, four spots appear on the paper, each representing one of the pigments. The bottom line is this: Pigments move at different rates according to their ability to *dissolve* in the solvent. The pigment that dissolves the best moves up the paper the fastest.

- The rate of migration of the pigments is calculated using an equation called the reference front ratio, R_f. The R_f value is the ratio of the pigment migration distance to the solvent migration distance.

$$R_f = \frac{\text{pigment migration}}{\text{solvent migration}}$$

EXERCISE 4B

In the second part of this lab, the rate of photosynthesis in chloroplasts was measured using a dye called DPIP. DPIP changes color when it accepts electrons that would normally be accepted by NADP⁺, the electron acceptor of the light-dependent reaction of photosynthesis. As DPIP accepts electrons, it changes from blue to clear. A gadget called a *spectrophotometer* is used to measure the amount of light absorbed.

What are the general concepts you *really* need to know?

- Three things can affect photosynthesis: light intensity, light wavelength, and temperature.

- In an experimental design, you can vary these three variables in order to determine their effects on photosynthetic rates.

LAB 5: CELL RESPIRATION

In this lab, the respiratory rate of germinating and nongerminating (dry) peas was investigated. The equation for cellular respiration is:

$$C_6H_{12}O_6 + 6O_2 + 6H_2O \rightarrow 6CO_2 + 12H_2O + \text{energy (ATP)}$$

What are the general concepts you *really* need to know?

- Cellular respiration is the process by which cells use oxygen to make energy (ATP).

- An instrument called a *respirometer* measures oxygen consumption during respiration.

- Germinating peas had a high respiratory rate, while nongerminating peas had a low respiratory rate.

LAB 6: MOLECULAR BIOLOGY

In this lab, the principles of genetic engineering were studied. This technology permits the manipulation of genetic material.

What are the general concepts you *really* need to know?

- This is one lab ETS loves to test. We already know that bacteria can accept segments of foreign DNA and incorporate them into their own chromosomes. In Chapter 9, we mentioned three ways in which bacteria accomplish this: *conjugation, transformation,* and *transduction.* In addition to these three, DNA can also be inserted into bacteria using *plasmids.* Plasmids are small circular DNA fragments that can be introduced into the bacteria *E. coli.* A plasmid can serve as a *vector*—a vehicle—to incorporate genes into the host's chromosome. Plasmids are key elements in genetic engineering.

- One way to incorporate specific genes into a plasmid is to use *restriction enzymes,* which cut the foreign DNA at specific sites, producing DNA fragments. A specific fragment can then be mixed together with a plasmid. The resulting recombinant plasmid is then taken up by *E. coli.*

- Plasmids can give a transformed cell a selective advantage. For example, if a plasmid carries genes that confer resistance to antibiotics, such as ampicillin, it can transfer these genes to the bacteria. This bacteria is then said to be transformed. That means if ampicillin is in the culture, only *transformed* cells will grow.

- Be familiar with the technique of *electrophoresis*. This technique is used in genetic engineering to separate and identify DNA fragments. First, the DNA is cut with various restriction enzymes. Then the fragments are "loaded" into wells on a special gel called an *agarose gel*. As electricity courses through the gel, the fragments move across it according to their molecular weights.

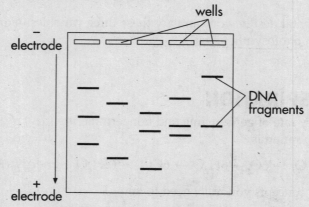

- In the end, the distance of each fragment is recorded. This technique is especially helpful in identifying a mutant gene for such diseases as sickle-cell anemia.

- The bottom line is this: You can use plasmids to transform nonresistant bacteria into resistant bacteria.

LAB 7: GENETICS OF *DROSOPHILA*

In this lab, the fruit fly, *Drosophila melanogaster*, was used to study genetic crosses in both F1 and F2 generations. Fruit flies that are homozygous for normal traits (also called wild-types) were crossed with mutant fruit flies bearing recessive traits. Mutations such as vestigial wings and white-colored eyes are examples of recessive traits.

What are the general concepts you *really* need to know?

- Review the discussion of genetics in this book. Know how to use a Punnett square to determine the results in both the F1 and F2 generations.

- Know how to apply this information to design other genetic experiments.

LAB 8: POPULATION GENETICS AND EVOLUTION

What are the general concepts you *really* need to know?

- Review the discussion of the Hardy-Weinberg principle in this book. Know the five conditions of the Hardy-Weinberg equilibrium: (1) a large population, (2) no mutations, (3) no immigration or emigration, (4) random mating, and (5) no natural selection.

- Know how to calculate the allele and genotype frequencies using the two Hardy-Weinberg equations: $p + q = 1$, and $p^2 + 2pq + q^2 = 1$. Don't forget: If the population obeys Hardy-Weinberg's rules, these frequencies remain constant over time.

- Review our discussion of evolution. Be familiar with how changes in allele frequencies can drive evolution.

LAB 9: TRANSPIRATION

In this lab, the movement of water within a plant was examined.

What are the general concepts you *really* need to know?

- This lab closely followed the review of transpiration in this book. Review how water moves up a plant via capillary action. Don't forget the important properties of water related to transpiration—hydrogen bonding, polarity, and adhesion and cohesion.

- Review the different vascular tissues that are involved in transport in plants, mainly the different types of xylem and phloem.

LAB 10: PHYSIOLOGY AND THE CIRCULATORY SYSTEM

Lab 10 had two parts.

EXERCISE 10A

In lab 10A, students were supposed to know how to measure blood pressure using a *sphygmomanometer*.

What are the general concepts you *really* need to know?

- Your heart rate changes with your body position. For example, your heart rate increases as you change from a reclining to a standing position.

- Your heart rate also increases when you exercise.

- A blood pressure reading consists of two numbers: a systolic pressure and diastolic pressure. Normal blood pressure for a healthy adult is about 120/80. The numerator is the systolic pressure and the denominator is the diastolic pressure.

EXERCISE 10B

In lab 10B, the Q10 value in water fleas, *Daphnia*, was evaluated. All animals generate heat in metabolism. Water fleas are ectotherms, meaning they get their heat from the environment. Ectotherms regulate their body temperature behaviorally. For example, if it's too hot, they move to a shady place.

What are the general concepts you really need to know?

- Fortunately, you don't have to know about water fleas, per se. What you do have to know is the concept of Q10. Q10 measures the increase in metabolic activity resulting from an increase in body temperature. Since *Daphnia* can adjust their temperature to the environment, as the temperature in the environment increases, their body temperature also increases. This in turn increases their heartbeat rate per minute. The Q10 was measured for water fleas at various temperatures.

$$Q_{10} = \frac{\text{the heartbeat at the higher temperature}}{\text{the heartbeat at the lower temperature}}$$

- All you have to know is that a warm environment can increase the body temperature and heart rate of cold-blooded animals.

LAB 11: HABITAT SELECTION

In this lab, the response of the brine shrimp, *Artemia*, to various habitats was investigated. This experiment looked at how brine shrimp select their environment based on light, heat, and pH. Crustaceans such as brine shrimp can survive in a broad range of water environments from low salinity to high salinity. However, brine shrimp usually prefer waters with a lot of salt because their predators can't survive in them. This is an example of natural selection.

What are the general concepts you really need to know?

- A number of variables, such as predation and the nature of the environment (in this case excessive salinity), can influence an organism's selection of habitat.

LAB 12: DISSOLVED OXYGEN AND AQUATIC PRIMARY PRODUCTIVITY

In this lab, the amount of oxygen dissolved in water samples was examined using the Winkler technique. The Winkler technique is a way of measuring the amount of dissolved oxygen in water.

What are the general concepts you really need to know?

- Gases such as carbon and oxygen are always recycled in our ecosystem.

- Three factors affect the solubility of dissolved gases in aquatic ecosystems: temperature, salinity, and photorespiration. Temperature and salinity decrease dissolved oxygen concentration. Photorespiration increases oxygen concentration.

- Know the concept of primary productivity. The rate at which producers in an ecosystem store energy is called the primary productivity. Primary productivity is determined by the rate at which oxygen is produced.

- Also know the difference between gross and net productivity. Gross productivity is the total energy obtained by organisms, whether through photosynthesis or ingestion. Net productivity is the rate at which energy is incorporated into bodies to produce growth.

16

Answers and Explanations to the Chapter Quizzes

CHAPTER 2: THE CHEMISTRY OF LIFE

1. **(A) is the correct answer.** Hydrogen bonds are weak-bond attractions between the positively polar hydrogen of one molecule and the negatively polar oxygen of another molecule.

2. **(C) is the correct answer.** Two monosaccharides (or simple sugars) are linked together by a glycosidic bond.

3. **(B) is the correct answer.** Two amino acids are held together by a peptide bond (CO-NH bond).

4. **(D) is the correct answer.** Fats are composed of three fatty acids and one glycerol molecule. The bond that holds together one of the fatty acids to the glycerol molecules is called an ester bond.

5. **(E) is the correct answer.** The amino group is a basic functional group ($-NH_2$).

6. **(D) is the correct answer.** There are only two major storage forms of carbohydrates: starch and glycogen. Starch is the storage form of sugars in plants. (E): Glycogen is the storage form of sugar in animals. (A): Cellulose is a structural component of the cell wall. (B) and (C): Maltose and fructose are monosaccharides.

7. **(D) is the correct answer.** A change of one pH equals a tenfold change in hydrogen ion concentration. Therefore, if the pH changes from 8 to 10 (a pH change of 2), the resulting solution is 100 times more basic.

8. **(D) is the correct answer.** In order for lactose to be hydrolyzed to glucose and galactose, a water molecule must be added. The other species O_2, H_2, ATP, and NADH will not hydrolyze lactose.

9. **(D) is the correct answer.** Glycine is a simple amino acid, not a polymer. (A), (B), and (E): Starch, cellulose, and glycogen are all polymers of carbohydrates. (C): A polypeptide is a polymer of amino acids.

10. **(E) is the correct answer.** A triglyceride (a fat) does not have a hydroxyl functional group. That's because the hydroxyl groups from the glycerol molecule are involved in forming the bond that makes the triglyceride. (A), (B), (C), and (D): All of these organic compounds have a hydroxyl group.

11. **(D) is the correct answer.** A polypeptide that consists of 90 amino acids has 89 peptide bonds. How do we know that? There is one peptide bond that holds one amino acid to another. If a molecule has three amino acids linked together then it has two peptide bonds. When you add water it will release two molecules of water. If a molecule has four amino acids linked together, it will release three molecules of water. See the trend? Now, if a polypeptide has 90 amino acids linked together, it will release 89 molecules of water.

12. **(E) is the correct answer.** Capillary action is the means by which water is pulled up from the roots to leaves. This process involves hydrogen bonding (which links polar molecules), cohesion, and adhesion. Water has a high heat capacity, not a low one.

CHAPTER 3: CELLS AND ORGANELLES

1. **(E) is the correct answer.** Protein synthesis involves the production of proteins. It begins in the nucleus where transcription occurs (the making of mRNA) and ends in the cytoplasm where the final protein is made. This is called translation. Translation requires a ribosome, which sometimes sits on the rough endoplasmic reticulum. When the appropriate amino acids are brought to the ribosome a polypeptide is formed. The Golgi bodies help to sort secreted proteins. The smooth endoplasmic reticulum is the site of lipid synthesis.

2. **(C) is the correct answer.** Bacterial cells have a cell wall made up of peptidoglycan while animal cells don't. (A) and (B): Both bacterial cells and animal cells have ribosomes and plasma membranes made of phospholipids. (D) and (E): Bacterial cells do not have nuclear membranes and vacuoles.

3. **(D) is the correct answer.** Active transport requires ATP because it involves movement of materials against a concentration gradient. Materials can be transported across a plasma membrane in a variety of ways. Passive transport (diffusion), osmosis, and facilitated transport do not require energy since the materials move down a concentration gradient.

4. **(B) is the correct answer.** Lysosomes contain digestive enzymes that destroy worn-out organelles.

5. **(C) is the correct answer.** The cell wall is a semi-rigid structure that gives support to the cell. It is usually found in plants and fungi.

6. **(A) is the correct answer.** The smooth ER is a channel inside the cytoplasm that is the site of lipid synthesis. Lipids are nonprotein substances.

7. **(D) is the correct answer.** Microtubules are polymers, made up of the protein tubulin, which participate in cellular division and movement and are found in cilia, flagella, and spindle fibers.

8. **(E) is the correct answer.** The nucleolus is the site at which rRNA is formed.

CHAPTER 4: CELLULAR METABOLISM

1. **(B) is the correct answer.** NADH is not reduced during lactic acid fermentation. NAD^+ is a product of lactic acid fermentation. (A), (C), (D), and (E): During lactic acid fermentation, your body does not get enough oxygen, lactate and ATP are produced, and NAD^+ is recycled.

2. **(C) is the correct answer.** (E): Spontaneous reactions are those that occur without the input of energy. However, that doesn't mean that they happen instantaneously. A spontaneous reaction could take days, or even years. (B) and (D): When a spontaneous reaction does occur, it releases energy, which can be used to do work. (A): That's why the products contain less energy than the reactants.

3. **(D) is the correct answer.** Enzymes are proteins. This means they are polymers of amino acids not carbohydrates. (A): Enzymes often work with coenzymes but these substances are not always needed. (B): Enzymes do not become hydrolyzed, that is, broken down in the presence of water during a chemical reaction. (D): They do operate under a narrow pH range. (C): They are not consumed in a reaction.

4. **(C) is the correct answer.** Don't forget that oxygen is the final acceptor of electrons as they are passed down the electron chain. These electrons then combine with oxygen to produce water. (A), (D), and (E): Electrons are brought to the electron transport chain by carriers such as $FADH_2$ and NADH. (B): Water is a by-product of the acceptance of electrons by free oxygen.

5. **(E) is the correct answer.** The activity of an enzyme is influenced by a number of factors: pH, temperature, and concentration of substrates and enzymes. Water concentration has no effect on the enzyme since water is chemically neutral.

6. **(A) is the correct answer.** Let's use Process of Elimination. The starting materials in the reaction are glucose, oxygen, and water. This reaction represents the complete oxidation of glucose. That is, aerobic respiration. This means you can eliminate photosynthesis (it produces glucose) and anaerobic respiration (it doesn't use oxygen). (C): Glycolysis is only part of aerobic respiration. It produces pyruvic acid. (D): Finally, fermentation produces lactic acid or ethanol, neither of which is represented in the equation above.

7. **(C) is the correct answer.** Glycolysis is not considered an aerobic process because it does not require oxygen. Glycolysis is therefore an anaerobic process. The Krebs cycle, formation of acetyl CoA, the electron transport chain, and oxidative phosphorylation are all stages within aerobic respiration that require oxygen.

8. **(C) is the correct answer.** Since CO_2, H_2O, and ATP are produced, the muscle cell undergoes aerobic respiration. Aerobic respiration occurs first in the cytoplasm then in the mitochondrion.

9. **(C) is the correct answer.** Prokaryotes carry out oxidative phosphorylation in their plasma membranes. (A) and (E): They do not contain membrane-bound organelles or a nuclear membrane. (B): The cell wall is not involved in oxidative phosphorylation. (D): Ribosomes are the sites of protein synthesis.

10. **(C) is the correct answer.** FAD^+ is both an electron and hydrogen carrier in the Krebs cycle. It shuttles electrons (along with some hydrogens) to the electron transport chain. When it is oxidized it becomes FAD^+.

11. **(B) is the correct answer.** There are two types of fermentation: alcoholic and lactic acid fermentation.

12. **(A) is the correct answer.** During glycolysis, the final product is pyruvic acid, which is also known as pyruvate.

13. **(D) is the correct answer.** During oxidative phosphorylation, ATP is produced when hydrogen ions cross a special protein channel called ATP synthase (also known as ATP synthetase).

14. **(E) is the correct answer.** Cofactors as well as coenzymes can assist enzymes during chemical reactions.

CHAPTER 5: PHOTOSYNTHESIS

1. **(E) is the correct answer.** Carbon dioxide is used to make glucose in the dark reaction. All of the other statements are true concerning the light reaction. (A): All pigments, including the antennae pigments, are capable of capturing sunlight. (B) and (D): These pigments send energy to

the reaction center (which in photosystem II is P680) where electrons are activated and eventually passed down an electron transport chain. (C): During photosynthesis, sunlight energy is converted to chemical energy.

2. **(A) is the correct answer.** As electrons are passed down the electron transport chain they eventually combine with an electron carrier to form NADPH. (B): ATP is produced in photo system II. (C): Cytochromes are carriers that are involved in passing electrons down a chain. (D): Water is not produced during the electron transport chain. (E): Glucose is the final product of photosynthesis. It is not produced during the light reaction.

3. **(C) is the correct answer.** During noncyclic phosphorylation, both ATP and NADPH are produced whereas in cyclic phosphorylation, only ATP is produced. (A) and (D): Light is absorbed and electrons are passed along the electron transport chain in both reactions.

4. **(B) is the correct answer.** Use Process of Elimination. Since sunlight is part of the reaction we can eliminate the electron transport system (E), Calvin cycle (D), and glycolysis (C). (A): In photosystem I, only sunlight and ADP is required. Photosystem II requires water since water is split during this reaction.

5. **(D) is the correct answer.** C_4 is an alternate pathway to produce glucose. It is common among tropical plants and results in a higher rate of photosynthesis than in C_3 plants. (A): There are no special pigments involved in the dark reaction in photosynthesis. (B): C_4 plants do not use oxygen instead of carbon to make glucose. (C): C_3 plants fix carbon days and nights. (E): This statement is incorrect because C_4 plants are better adapted to intense sunlight.

6. **(A) is the correct answer.** The dark reaction and its enzymes are found in the stroma of the leaf.

7. **(B) is the correct answer.** The thylakoids are the site of photophosphorylation.

8. **(E) is the correct answer.** Ribulose bisphosphate is a 5-carbon molecule that combines with carbon dioxide in the Calvin cycle.

9. **(E) is the correct answer.** Ribulose bisphosphate is the molecule that accepts carbon dioxide during the Calvin cycle. Don't forget you can use an answer choice more than once.

CHAPTER 6: MOLECULAR GENETICS

1. **(B) is the correct answer.** Transcription involves the production of mRNA. It occurs in the nucleus. (A) and (E) are both incorrect because the mitochondria and lysosomes are organelles that have nothing to do with translation. (C) and (D): Ribosomes and the Golgi apparatus are both involved in translation—the production of proteins.

2. **(B) is the correct answer.** Use Process of Elimination. A phenotypic expression refers to traits. The sequence must therefore end with trait. That means we can eliminate (D) and (E). The question begins with genes, which are segments of a chromosome. The sequence must therefore begin with DNA.

3. **(E) is the correct answer.** RNA is a nucleic acid that consists of a phosphate group, 5-carbon sugar (deoxyribose), and a nitrogenous base. RNA may contain adenine, guanine, cytosine, or uracil instead of thymine and ribose sugar.

4. **(C) is the correct answer.** If an mRNA codon is UAC, the complementary segment on a tRNA anticodon is AUG.

5. **(D) is the correct answer.** During posttranslational modification, the polypeptide undergoes a conformational (secondary, tertiary, or quaternary structure). (A) and (B) are incorrect because they refer to posttranscriptional modification—changes that occur once the mRNA molecule is formed. (C) is incorrect because formation of peptide bonds occur during translation. (E) is incorrect because amino acids are not made during translation.

6. **(C) is the correct answer.** RNA polymerase does not participate in DNA replication. It is used during RNA synthesis to bring RNA nucleotides to the sense strand in DNA. (A): DNA helicase is used to unwind the double helix during DNA replication. (B): DNA polymerase is the enzyme that brings DNA nucleotides to each DNA strand. (D): RNA primase is the enzyme that initiates the DNA process. (E): DNA ligase links the Okazaki fragments during DNA replication.

7. **(B) is the correct answer.** Since three nucleotides make up one amino acid, seven amino acids would be incorporated into a polypeptide.

8. **(B) is the correct answer.** Transposons are DNA segments that move around.

9. **(C) is the correct answer.** The lagging strand is assembled in discrete nucleotide segments known as Okazaki fragments. This strand is made discontinuously.

10. **(C) is the correct answer.** Notice how the same answer can be used twice: two slightly different questions refer to the same choice, the lagging strand.

11. **(E) is the correct answer.** hnRNA is the precursor of mRNA. The discontinuous stretches of DNA are the Okazaki fragments.

CHAPTER 7: CELL REPRODUCTION

1. **(B) is the correct answer.** A diploid cell contains twice the number of chromosomes as a haploid cell. If the diploid number is 24 then the haploid number is 12.

2. **(B) is the correct answer.** During the first half of meiosis I, the homologous chromosomes lie adjacent to each other. This is called synapsis. (A): Crossing-over refers to the exchange of segments of DNA. (C): A tetrad refers to two homologous chromosomes which are attached to each other. (D): Cytokinesis refers to the distribution of cytoplasm to the two daughter cells. (E): Interphase is the stage during which chromosomes replicate.

3. **(E) is the correct answer.** During meiosis, you have two rounds of all the same stages as mitosis except interphase. Why? Because the chromosomes do not replicate again.

4. **(B) is the correct answer.** Sister chromatids disjoin (separate) during meiosis II. All other events occur during meiosis I.

5. **(B) is the correct answer.** The genetic material is called chromatin during interphase. It is not visible.

6. **(E) is the correct answer.** Chromosomes are replicated during the S phase of the cell cycle.

7. **(C) is the correct answer.** The centromere is the structure that holds the chromatids together.

8. **(D) is the correct answer.** Centroles are contained within microtubule organizing centers.

9. **(A) is the correct answer.** During G2 more proteins are made.

CHAPTER 8: HEREDITY

1. **(C) is the correct answer.** Hybrids are organisms that have one of each allele, Ss. When two hybrids cross the offspring are SS, Ss, Ss, and ss. Therefore the percent of the offspring that will possess the same genotype as their parents is 50 percent.

2. **(B) is the correct answer.** This question is a bit tricky. In the chapter we discussed two independently assorting traits, such as AB or ab. This question is asking about three independently assorting traits. The first thing to do is to write out the gametes that are possible. They are: ABC, ABc, AbC, Abc, aBC, aBc, abC, and abc. Notice that only one out of the eight gametes is recessive. Another way to do this problem is to use this equation: $x = 2^n$, where n represents the number of independently assorting traits. In our example, for three traits, there are $2^3 =$ 8 gametes produced. For four independently assorting traits, there are $2^4 =16$ gametes. In all of these examples, the number of gametes that are recessive is always 1. Therefore, for three independently assorting traits, 1 out of 8 is recessive.

3. **(E) is the correct answer.** Sex-linked traits are traits that almost always exist on the X chromosome. They are therefore often passed from mother to son since sons must receive the X chromosome from their mothers. (A): Many traits can skip a generation. (B): Some diseases that have carriers are not sex-linked. (C): Just because a trait appears in all the offspring, it doesn't necessarily mean that the trait is sex-linked. (D): Sex-linked traits do not necessarily have to be passed from mothers to daughters. Daughters can inherit a good X from their fathers.

4. **(C) is the correct answer.** The total number of offspring produced is 400. The number of offspring that exhibit the recessive trait is 81. This means that roughly 25 percent of the offspring show the recessive trait. Do a Punnett square to determine the genotype of the parents: Bb × Bb = BB, Bb, Bb, and bb. This would yield phenotypes in approximately the proportions described above (3:1).

5. **(A) is the correct answer.** Use the product rule to solve this problem: (1/2)(1/2)(1/2) = 1/8.

6. **(C) is the correct answer.** When two different alleles are observed in regard to a trait the organism is a heterozygote.

7. **(A) is the correct answer.** The physical appearance of an organism is called the phenotype. In contrast, the genetic makeup is called the genotype.

8. **(B) is the correct answer.** When two different alleles are both expressed this is an example of codominance.

9. **(D) is the correct answer.** A dihybrid cross is a cross that involves two traits that are independently assorting, such as tall, green pea plants versus short, yellow pea plants.

CHAPTER 9: CLASSIFICATION

1. **(A) is the correct answer.** Use Process of Elimination. This organism has a nucleus so we can eliminate bacteria (E). It has a cell wall so we can eliminate animals (D). It is photosynthetic so we can eliminate fungi (B). It is unicellular. Therefore, we can eliminate plants (C).

2. **(D) is the correct answer.** Arthropods have the following characteristics: They have segmented bodies, a chitinous exoskeleton, jointed appendages, and an open circulatory system. (A) and (B): Only insects have wings.

3. **(A) is the correct answer.** Once again, use Process of Elimination to arrive at your answer. The organism is unicellular, so we can eliminate animals and plants, (D) and (E). The organism is a eukaryote so we can eliminate Monera, (B). It has threadlike branches and a chitinous cell wall—it's a fungus.

4. **(D) is the correct answer.** Review the excretory systems of animals. Earthworms excrete nitrogenous waste via nephridia. Insects use special structures call Malpighian tubules. Fish and amphibians have kidneys. Flatworms have flame cells. Notice we didn't mention flame cells in this chapter but you could still arrive at the right answer using Process of Elimination.

5. **(B) is the correct answer.** Review the characteristics of vertebrates. Vertebrates include the following organisms: fish, amphibians, reptiles, birds, and mammals. Arthropods are invertebrates.

6. **(C) is the correct answer.** Fungi are not photosynthetic. Think about your average mushroom. They live off dead organisms. Fungi are eukaryotes that reproduce sexually (the fusion of gametes) and asexually (via spores). They require oxygen and have cell walls.

7. **(A) is the correct answer.** This is an example of a mutualistic relationship. Both organisms benefit. (C): Commensalism means that one organism benefits and one is unaffected. (B): Parasitism means that one organism benefits and the other is harmed. (E): Tropism is a type of behavior in plants such as phototropism. (D): Competition occurs when two individuals or species compete for the same resources.

8. **(C) is the correct answer.** Viruses are microorganisms that contain nucleic acid (either DNA or RNA) and a protein coat. They do not contain a nuclear membrane, a cell wall, or organelles.

9. **(A) is the correct answer.** Conjugation is the exchange of genetic material via a pilus.

10. **(B) is the correct answer.** When a virus transfers DNA from one bacterium to another this is called transduction.

11. **(D) is the correct answer.** A plasmid is a small circular DNA that carries genes separate from the main bacterial DNA.

12. **(C) is the correct answer.** Transformation involves the incorporation of naked DNA segments.

CHAPTER 10: PLANTS

1. **(D) is the correct answer.** Use Process of Elimination. Dicots are angiosperms that have two cotyledons, netted veins, and flower parts in multiples of four and five. We can therefore eliminate (B) and (C). Since dicots are flowering plants they have vascular tissue. We can therefore eliminate (A). The cotyledon provides nutrition for the plant so we can eliminate (E).

2. **(D) is the correct answer.** Conifers, which are gymnosperms, and flowering plants, which are angiosperms, share the following characteristics: They both contain enclosed seeds. The other characteristics listed are only true for gymnosperms. Gymnosperms are perennials (live year after year), contain xylem that is dead (the rings of trees), and are deciduous (they have leaves that shed). Only some angiosperms are woody plants.

3. **(D) is the correct answer.** The anther is the male part of the flower. The pistil is the female part of flowering plants. It consists of the ovule, style, ovary (which contains the female gametes), and the stigma (the sticky portion that traps pollen grains).

4. **(D) is the correct answer.** The plant tissue that gives girth to a plant is called the lateral meristem. It is subdivided into the vascular cambium and the cork cambium.

5. **(C) is the correct answer.** Carnivorous plants are capable of obtaining proteins by trapping flying insects. They therefore would thrive in nitrogen-poor environments.

6. **(B) is the correct answer.** Tracheophytes are vascular plants. This means they include trees, grass, corn, and beans. Mosses are bryophytes; they lack true stems and roots.

7. **(D) is the correct answer.** Ethylene is the hormone responsible for ripening fruits. (A): Auxins promote growth in plants and cause them to bend toward light. (C): Gibberellins cause dwarf plants to grow. (E): Abscisic acids inhibit leaves from falling and promote bud and seed dormancy. (B): Cytokinins promote cell division and differentiation.

8. **(B) is the correct answer.** The bending of plants toward light is called phototropism.

9. **(D) is the correct answer.** When a Venus flytrap responds to touch, it exhibits thigmotropism.

10. **(A) is the correct answer.** The pigment phytochrome is involved in regulating sexual reproduction in plants.

11. **(E) is the correct answer.** Gravitropism refers to growth in plants in response to gravity

12. **(C) is the correct answer.** Photoperiodism is the process by which flowering plants are affected by day length.

CHAPTER 11: ANIMAL STRUCTURE AND FUNCTION

1. **(C) is the correct answer.** There are four blood types: A, B, AB, and O. Blood type O is the universal donor and AB is the universal recipient. Individuals with blood type AB can receive blood from all of the other blood groups without any blood clotting.

2. **(B) is the correct answer.** Within the small intestine there are lacteals whose job is to absorb digested fats. (A): The nephrons are the excretory units in the kidneys. (C): Lacteals are found within villi. (D): Root hairs are tiny structures that absorb water and minerals in plants. They also increase the surface area for absorption. (E): Hormones are substances that travel in the blood.

3. **(D) is the correct answer.** Oxytocin is the hormone that causes uterine contractions during labor. (A) and (B): FSH and LH are hormones that are involved in the menstrual cycle. (C): Prolactin promotes the production of milk in mammary glands. (E): Epinephrine is the hormone involved in flight/fight responses.

4. **(E) is the correct answer.** Use Process of Elimination. We know that immune cells are made in the bone marrow so we can eliminate (B). We also know that immune cells are found in the lymph vessels so eliminate (C). T-cells mature in the thymus so we can eliminate (A). If you're not sure about the spleen skip it and look at answer choice (E). The kidney has nothing to do with the immune system. Therefore the answer is (E). By the way, the spleen also contains lymphocytes. Notice that you didn't have to know that to get the question right.

5. **(B) is the correct answer.** Use Process of Elimination again. Smooth muscle cells are slow contracting so eliminate answer choices (A), (C), and (E). They are under involuntary control so the answer must be (B).

6. **(B) is the correct answer.** The enzyme that initially breaks down protein in the stomach is pepsin. (A): Bile is not an enzyme; it emulsifies fats. (C): Trypsin breaks down proteins later on; it is active in the small intestine. (D) and (E): Salivary amylase and pancreatic amylase break down carbohydrates.

7. **(D) is the correct answer.** The hypothalamus is the part of the brain that regulates the pituitary gland and produces neurosecretory hormones. (A) and (B): TSH and LH are hormones secreted by the anterior pituitary. (C): The hypothalamus is not an extension of the pituitary gland. (E): It secretes hormones, not neurotransmitters.

8. **(B) is the correct answer.** Enzymes that digest fats are called lipase. (A): Bile isn't an enzyme; it emulsifies fats. (C) and (E): Trypsin digests proteins in the small intestine and amylase digests carbohydrates. Now you're down to two answer choices. You may not know what proteases are but you know that lipase digests fats. By the way, proteases are enzymes that digest proteins. Again, you didn't have to know that for this question.

9. **(B) is the correct answer.** Diabetics have too much sugar in their blood. This means they are not storing the sugar. Some diabetics lack insulin, which is why they take insulin shots. (A): Glucagon is the hormone antagonistic to insulin. (C): Parathyroid hormone regulates calcium in the blood. (D): Calcitonin lowers blood calcium levels. (E): Norepinephrine regulates fight/flight responses.

10. **(E) is the correct answer.** Carbon dioxide can travel in the bloodstream in many forms. It can mix with water and form carbonic acid. It can also attach to hemoglobin in red blood cells. (E): Platelets, on the other hand, are cell fragments that are involved in blood clotting.

11. **(C) is the correct answer.** The extraembryonic membrane that stores wastes is the allantois. (A): The chorion is the outermost membrane that surrounds all the other membranes. (B): The amnion forms a sac that protects the embryo. (D) and (E): The yolk sac provides food for the embryo and the eggshell is the hard covering.

12. **(C) is the correct answer.** Blood leaves the right ventricle and enters the lungs via the pulmonary arteries.

13. **(A) is the correct answer.** When blood returns from the body it enters the right atrium via the vena cava.

14. **(E) is the correct answer.** The vena cava is the vessel that sends blood to the right atrium.

15. **(E) is the correct answer.** The germ line cells are also called gametes.

16. **(B) is the correct answer.** The blastocoel in the fluid-filled cavity that forms during blastula.

17. **(A) is the correct answer.** When a sperm and an egg unite they form a zygote.

18. **(C) is the correct answer.** Morula is the first stage after cleavage during morphogenesis.

19. **(D) is the correct answer.** The pancreatic duct carries trypsin and chymotrypsin, among other enzymes, to the small intestine.

20. **(C) is the correct answer.** Bile is made in the liver and stored in the gall bladder.

21. **(B) is the correct answer.** Complete digestion of food occurs in the small intestine.

22. **(E) is the correct answer.** The function of the large intestine is to reabsorb water.

23. **(C) is the correct answer.** Ligaments connect bone to bone.

24. **(B) is the correct answer.** Tendons connect muscle to bone.

25. **(A) is the correct answer.** Cartilage is the embryonic tissue in the skeletal system.

26. **(E) is the correct answer.** Bone is made up of collagen (a protein) and calcium (mineral). Calcium is the mineral that hardens the bones.

CHAPTER 12: EVOLUTION

1. **(C) is the correct answer.** Genetic variability is usually due to a mutation or meiosis (sexual reproduction). Answer choices (A), (B), (D), and (E) are all examples of asexual reproduction. They create identical offspring.

2. **(B) is the correct answer.** These marsupials were exposed to the same environmental conditions and developed similar structures (pouches) because of it. This is an example of convergent evolution. (A): Divergent evolution refers to two species that branched out from a common ancestor. (D): Genetic variability is too broad a term for this question. (C): Disruptive selection refers to a situation in which extreme phenotypes are favored in a population.

3. **(C) is the correct answer.** We mentioned that one way to get genetic variability is by mutations. Mutations are not irreversible. They can revert (or back mutate). (B) and (E): Mutations are not common (there are low rates) in a population. (D): Mutations can influence any gene locus in a population.

4. **(A) is the correct answer.** When the extremes are eliminated in a population this is called stabilizing selection.

5. **(E) is the correct answer.** When the extremes are favored over the intermediates this is called disruptive selection.

6. **(A) is the correct answer.** When the intermediates are favored over the extreme phenotypes this is once again called stabilizing selection.

7. **(B) is the correct answer.** When one extreme is favored in a population this is called directional selection.

8. **(D) is the correct answer.** When organisms are isolated from other members of their species by a geographical barrier, it is called allopatric speciation.

9. **(A) is the correct answer.** This is an example of use and disuse. The muscle mass increases because of exercise.

10. **(B) is the correct answer.** When a change occurs in a population that confers a benefit on some members of that population, increasing their likelihood of survival, this is called natural selection.

11. **(D) is the correct answer.** The study of the geographical distribution of living organisms is called biogeography.

12. **(E) is the correct answer.** This is the study of life based on fossils and footprints.

CHAPTER 13: BEHAVIOR AND ECOLOGY

1. **(C) is the correct answer.** Gravitropism refers to a plant's response to gravity. (A): Hydrotropism refers to a plant's response to water. (B): Chemotropism refers to a plant's response to a chemical substance. (D): Thigmotropism refers to a plant's response to touch. (E): Circadian rhythm refers to a behavioral rhythm with a period of about 24 hours.

2. **(E) is the correct answer.** The two systems that regulate behavior and how we respond to our environment are the nervous system (that's our immediate response) and the endocrine system (that's our slow response). The excretory, reproductive, and digestive systems do not directly regulate any behavioral responses.

3. **(C) is the correct answer.** Plants are producers and can convert light energy to chemical energy.

4. **(A) is the correct answer.** The animal with the smallest biomass is at the top of the pyramid. Therefore birds have the smallest biomass.

5. **(E) is the correct answer.** A secondary consumer feeds on a primary consumer, which feeds on producers. Based on the food chain above the secondary consumers are mice.

6. **(B) is the correct answer.** The part of the earth in which life exists is called the biosphere.

7. **(B) is the correct answer.** Fungi and bacteria both serve as decomposers. They break down organic matter. Viruses invade other organisms but they're not decomposers. Protists are unicellular organisms such as paramecium and euglena. They're not decomposers either.

8. **(A) is the correct answer.** A stickleback fish will attack another fish if it has a red belly (or if an object has a similar red mark). This is an act of aggression that is innate.

9. **(B) is the correct answer.** A blue jay avoids eating a particular type of butterfly because it "remembers" the initial unpleasant taste it had from a previous exposure.

10. **(E) is the correct answer.** The chimpanzee has figured out on its own how to get those bananas from the ceiling. This is called insight learning.

11. **(B) is the correct answer.** The biome at the highest altitude is the tundra—the cold, treeless region.

12. **(A) is the correct answer.** The biome with evergreen trees (conifers) is the taiga.

13. **(B) is the correct answer.** This biome has little vegetation (mainly grass).

14. **(D) is the correct answer.** This biome has the most diverse plant life.

17

The Princeton Review
AP Biology
Practice Test 1

BIOLOGY

Three hours are allotted for this examination: 1 hour and 20 minutes for Section I, which consists of multiple-choice questions, and 1 hour and 40 minutes for Section II, which consists of essay questions.

SECTION I

Time—1 hour and 20 minutes
Number of questions—100
Percent of total grade—60

Section I of this examination contains 100 multiple-choice questions, followed by 15 multiple-choice questions regarding your preparation for this exam. Please be careful to fill in only the ovals that are preceded by numbers 1 through 115 on your answer sheet.

General Instructions

INDICATE ALL YOUR ANSWERS TO QUESTIONS IN SECTION I ON THE SEPARATE ANSWER SHEET ENCLOSED. No credit will be given for anything written in this examination booklet, but you may use the booklet for notes or scratchwork. After you have decided which of the suggested answers is best, COMPLETELY fill in the corresponding oval on the answer sheet. Give only one answer to each question. If you change an answer, be sure that the previous mark is erased completely.

Example: Sample Answer

Chicago is a

(A) state
(B) city
(C) country
(D) continent
(E) village

Many candidates wonder whether or not to guess the answer to questions about which they are not certain. In this section of the examination, as a correction for haphazard guessing, one-fourth of the number of questions you answer incorrectly will be subtracted from the number of questions you answer correctly. It is improbable, therefore, that mere guessing will improve your score significantly; it may even lower your score, and it does take time. If, however, you are not sure of the correct answer but have some knowledge of the question and are able to eliminate one or more of the answer choices as wrong, your chance of getting the right answer is improved, and it may be to your advantage to answer such a question.

Use your time effectively, working as rapidly as you can without losing accuracy. Do not spend too much time on questions that are too difficult. Go on to other questions and come back to the difficult ones later if you have time. It is not expected that everyone will be able to answer all the multiple-choice questions.

BIOLOGY

SECTION I

Time—1 hour and 20 minutes

<u>Directions:</u> Each of the questions or incomplete statements below is followed by five suggested answers or completions. Select the one that is best in each case and then fill in the corresponding oval on the answer sheet.

1. The resting membrane potential depends on which of the following?

   I. Active transport
   II. Selective permeability
   III. Differential distribution of ions across the axonal membrane

   (A) II only
   (B) III only
   (C) I and II only
   (D) II and III only
   (E) I, II, and III

2. The Krebs cycle in humans occurs in the

   (A) cytoplasm
   (B) mitochondrial matrix
   (C) inner mitochondrial membrane
   (D) outer mitochondrial membrane
   (E) intermembrane space

3. A botanist travels to a mountainous, tropical country to participate in an expedition to study the fauna on one of that country's highest peaks. As he travels up the mountain, what are the principal terrestrial biomes he will most likely encounter?

   (A) Tropical rain forest-temperate deciduous forests-taiga-tundra
   (B) Tropical rain forest-tundra-temperate deciduous forest-taiga
   (C) Temperate deciduous forest-tropical rain forest-tundra-taiga
   (D) Tropical rain forest-temperate deciduous forest-tundra-taiga
   (E) Tropical rain forest-taiga-temperate deciduous forest-tundra

4. Regarding meiosis and mitosis, one difference between the two forms of cellular reproduction is that in meiosis

   (A) there is one round of cell division whereas in mitosis there are two rounds of cell division
   (B) separation of sister chromatids occurs during the second division whereas in mitosis separation of sister chromatids occurs during the first division
   (C) chromosomes are replicated during interphase whereas in mitosis chromosomes are replicated during prophase
   (D) spindle fibers form during prophase whereas in mitosis the spindle fibers form during metaphase
   (E) there is a reduction of the chromosome number whereas in mitosis there is an increase in the chromosome number

5. A feature of amino acids NOT found in carbohydrates is the presence of

   (A) carbon
   (B) oxygen
   (C) nitrogen
   (D) hydrogen
   (E) phosphorus

6. Which of the following is NOT a characteristic of bacteria?

   (A) Circular double-stranded DNA
   (B) Membrane-bound cellular organelles
   (C) Plasma membrane consisting of lipids and proteins
   (D) Ribosomes that synthesize polypeptides
   (E) Cell wall made of peptidoglycan

GO ON TO THE NEXT PAGE

7. Which of the following offers the best reasoning that the population is the evolving unit?

(A) Genetic changes can only occur at the population level.
(B) The gene pool in a population remains fixed over time.
(C) Natural selection affects individuals not populations.
(D) Individuals cannot evolve but populations can.
(E) Most changes in a population's gene pool do not result in evolution.

8. The part of the brain that controls involuntary actions is known as the

(A) cerebellum
(B) cerebrum
(C) hypothalamus
(D) medulla
(E) thalamus

9. In woody dicots, primary xylem and phloem cells are replaced by secondary xylem and phloem cells that arise from the

(A) apical meristem
(B) epidermis
(C) vascular cambium
(D) cork cambium
(E) lenticels

10. A scientist carries out a cross between two guinea pigs, both of which have black coats. Black hair coat is dominant over white hair coat. Three quarters of the offspring have black coats and one quarter have white coats. The genotypes of the parents were most likely

(A) bb × bb
(B) Bb × Bb
(C) Bb × bb
(D) BB × Bb
(E) BB × bb

11. A large island is devastated by a volcanic eruption. Most of the horses die except for the heaviest males and heaviest females of the group. They survive, reproduce, and perpetuate the population. Since weight is highly heritable and the distribution of weights approximates a binomial distribution, the offspring of the next generation would be expected to have

(A) a higher mean weight compared with their parents
(B) a lower mean weight compared with their parents
(C) the same mean weight as members of the original population
(D) a higher mean weight compared with members of the original population
(E) a lower mean weight compared with members of the original population

12. Which of the following represents the correct sequence of events in embryonic development?

(A) Blastula-cleavage-morula-gastrula-neurula
(B) Blastula-neurula-cleavage-morula-gastrula
(C) Cleavage-morula-blastula-neurula-gastrula
(D) Cleavage-morula-blastula-gastrula-neurula
(E) Morula-blastula-gastrula-cleavage-neurula

13. During the period when life is believed to have begun, the atmosphere on primitive Earth contained abundant amounts of all the following gases EXCEPT

(A) oxygen
(B) hydrogen
(C) ammonia
(D) methane
(E) water

14. A woman with blood genotype $I^A i$ and a man with blood genotype $I^B i$ have two children, both type AB. What is the probability that a third child will be blood type AB?

(A) $\dfrac{3}{4}$
(B) $\dfrac{1}{2}$
(C) $\dfrac{1}{4}$
(D) $\dfrac{1}{8}$
(E) 0

GO ON TO THE NEXT PAGE

15. The trophic level efficiency of large herbivores such as elks is frequently only about 5 percent. Which of the following is the most likely explanation for this low efficiency?

   (A) Elks have a lower rate of consumption than do other herbivores.
   (B) The digestive systems of large herbivores are not entirely efficient.
   (C) Large herbivores expend greater amounts of energy in respiration during their search for food than do carnivores.
   (D) Consuming producers is an inefficient way of obtaining energy.
   (E) Large herbivores lose more body heat than other herbivores.

16. An organism that is more complex than a platyhelminthes but more primitive than an arthropod is most likely to be

   (A) an annelid
   (B) an echinoderm
   (C) a chordate
   (D) a sponge
   (E) a coelenterate

17. Which of the following does NOT occur during the complete digestion of nutrients in humans?

   (A) Pancreatic lipase breaks down fats to fatty acids and glycerol.
   (B) Pepsin breaks down proteins to amino acids.
   (C) Pancreatic amylase breaks down carbohydrates into simple sugars.
   (D) Bile emulsifies fats into smaller fat particles.
   (E) Digested food is absorbed by capillaries in the villi.

18. In animal cells, which of the following represents the most likely pathway that a secretory protein takes as it is synthesized in a cell?

   (A) Plasma membrane-Golgi apparatus-ribosome-rough ER-secretory vesicle
   (B) Plasma membrane-Golgi apparatus-ribosome-secretory vesicle-rough ER
   (C) Ribosome-Golgi apparatus-rough ER-secretory vesicle-plasma membrane
   (D) Plasma membrane-Golgi apparatus-ribosome-secretory vesicle-rough ER
   (E) Ribosome-rough ER-Golgi apparatus-secretory vesicle-plasma membrane

19. All of the following statements are correct regarding alleles EXCEPT:

   (A) Alleles are alternative forms of the same gene.
   (B) Alleles are found on corresponding loci of homologous chromosomes.
   (C) A gene can have more than two alleles.
   (D) One allele can be dominant and the other recessive.
   (E) An individual with two identical alleles is said to be heterozygous with respect to that gene.

20. Once a plasmid has incorporated specific genes, such as the gene coding for the antibiotic ampicillin, into its genome, the plasmid may be cloned by

   (A) inserting it into a virus to generate multiple copies
   (B) treating it with a restriction enzyme in order to cut the molecule into small pieces
   (C) inserting it into a suitable bacterium in order to produce multiple copies
   (D) running it on a gel electrophoresis in order to determine the size of the gene of interest
   (E) infecting it with a mutant cell in order to incorporate the gene of interest

21. Although mutations occur at a regular and predictable rate, which of the following statements is the LEAST likely reason the frequency of mutation appears to be low?

   (A) Some mutations produce alleles that are recessive and may not be expressed.
   (B) Some undesirable phenotypic traits may be prevented from reproducing.
   (C) Some mutations cause such drastic phenotypic changes that they are removed from the gene pool.
   (D) The predictable rate of mutation results in ongoing variability in a gene pool.
   (E) The predictable rate of mutation is offset by a small rate of back mutation.

GO ON TO THE NEXT PAGE

22. Which of the following adaptive features would be found in flowering plants that live in an arid climate?

(A) Vascular tissues
(B) Stomates
(C) Thick cuticles
(D) Multiple leaves
(E) Megaspores

23. Which of the following best accounts for the ability of legumes to grow well in nitrogen-poor soils?

(A) These plants make their own proteins.
(B) These plants have a mutualistic relationship with nitrogen-fixing bacteria.
(C) These plants are capable of directly converting nitrogen gas into nitrates.
(D) These plants do not require nitrogen to make plant proteins.
(E) These plants have developed nitrogen-absorbing root hairs.

24. In a mammalian embryo, the mesoderm gives rise to

(A) the brain
(B) the nerves
(C) the skin
(D) the eyes
(E) the ovaries

Question 25 refers to the diagram below.

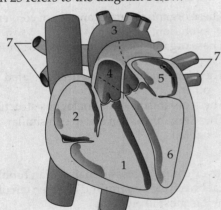

25. Which of the following chambers or vessels carry deoxygenated blood in the human heart?

(A) 4 only
(B) 1 and 2 only
(C) 5 only
(D) 1, 2 and 4
(E) 3, 5, and 6

26. In chick embryos, the extraembryonic membrane that provides nourishment to the fetus is the

(A) amnion
(B) chorion
(C) placenta
(D) ovary
(E) egg yolk

27. Some strains of viruses can change normal mammalian cells into cancer cells in vitro. This transformation of the mammalian cell is usually associated with the

(A) formation of a pilus between the mammalian cell and the virus
(B) incorporation of the viral genome into the mammalian cell's nuclear DNA
(C) conversion of the host's genome into the viral DNA
(D) release of spores into the mammalian cell
(E) incorporation of free-floating DNA from the environment

28. The major difference between cartilage and bone is that cartilage

(A) is a part of the skeletal system
(B) is composed of collagen and salts
(C) lacks blood vessels and nerves
(D) secretes a matrix
(E) is a type of connective tissue

29. All of the following are examples of events that can prevent interspecific breeding EXCEPT:

(A) The potential mates do not meet.
(B) The potential mates experience behavioral isolation.
(C) The potential mates have different courtship rituals.
(D) The potential mates have similar breeding seasons.
(E) The gametes of potential mates have biochemical differences.

GO ON TO THE NEXT PAGE

30. Which of the following is NOT a characteristic of asexual reproduction in animals?

  (A) Daughter cells have the same number of chromosomes as the parent cell.
  (B) Daughter cells are identical to the parent cell.
  (C) The parent cell produces diploid cells.
  (D) The daughter cells fuse to form a zygote.
  (E) The chromosomes replicate during interphase.

31. Which of the following is a characteristic of arteries?

  (A) They are thin-walled blood vessels.
  (B) They contain valves that prevent backflow.
  (C) They always carry oxygenated blood.
  (D) They carry blood away from the heart.
  (E) Blood is kept moving by the contraction of voluntary muscles.

32. Plant hormones that promote elongation and the development of buds and fruits are

  (A) cytokinins
  (B) auxins
  (C) gibberellins
  (D) abscisic acids
  (E) ethylenes

33. In angiosperms, when the ovule is fertilized, the zygote develops into an embryonic plant. During this process, which part of the plant becomes a seed?

  (A) The ovule
  (B) The ovary
  (C) The pollen grains
  (D) The cotyledons
  (E) The fertilized egg

34. If the genotype frequencies of an insect population are AA = 0.49, Aa = 0.42, and aa = 0.09, what is the gene frequency of the dominant allele?

  (A) 0.07
  (B) 0.30
  (C) 0.49
  (D) 0.50
  (E) 0.70

35. Crossing-over occurs during which of the following phases in meiosis?

  (A) Prophase I
  (B) Metaphase I
  (C) Anaphase I
  (D) Prophase II
  (E) Metaphase II

36. Which of the following statements about sexual reproduction in flowering plants is NOT true?

  (A) Meiosis takes place in the ovules.
  (B) A megaspore becomes the female gametophyte.
  (C) A pollen tube grows down the style and into the ovary.
  (D) Microspores are produced in the stamen.
  (E) The triploid cell becomes the endosperm, which nourishes the embryo.

37. Bryophytes generally differ from tracheophytes in that bryophytes have

  (A) a protective layer around their gametes
  (B) conducting tissues
  (C) stomates in leaf surfaces
  (D) waxy cuticles on their outer surfaces
  (E) water-borne motile sperm

38. In most ecosystems, net primary productivity is important because it represents the

  (A) energy available to producers
  (B) total solar energy converted to chemical energy by producers
  (C) biomass of all producers
  (D) energy used in respiration by heterotrophs
  (E) surplus energy generated by producers

39. Hawkmoths are insects that are similar in appearance and behavior to hummingbirds. Which of the following is LEAST valid?

  (A) These organisms are examples of convergent evolution.
  (B) These organisms were subjected to similar environmental conditions.
  (C) These organisms are genetically related to each other.
  (D) These organisms have analogous structures.
  (E) These organisms can survive in similar habitats.

40. If an invertebrate possesses malpighian tubules, a tracheal breathing system, and an open circulatory system it is most likely to be

  (A) a snail
  (B) a sponge
  (C) a butterfly
  (D) an earthworm
  (E) a flatworm

GO ON TO THE NEXT PAGE

41. Destruction of all beta cells in the pancreas will cause which of the following to occur?

    (A) Glucagon secretion will stop and blood glucose levels will increase.
    (B) Glucagon secretion will stop and blood glucose levels will decrease.
    (C) Glucagon secretion will stop and digestive enzymes will be secreted.
    (D) Insulin secretion will stop and blood glucose levels will increase.
    (E) Insulin secretion will stop and blood glucose levels will decrease.

42. All of the following are stimulated by the sympathetic nervous system EXCEPT:

    (A) dilation of the pupil of the eye
    (B) constriction of blood vessels
    (C) increased secretion of the sweat glands
    (D) increased peristalsis in the gastrointestinal tract
    (E) increased heart rate

43. The calypso orchid, *Calypso bulbosa*, grows in close association with mycorrhizae fungi. The fungi penetrate the roots of the flower and take advantage of the plant's food resources. The fungi concentrate rare minerals, such as phosphates, in the roots and make them readily accessible to the orchid. This situation is an example of

    (A) parasitism
    (B) commensalism
    (C) mutualism
    (D) endosymbiosis
    (E) altruism

44. Which of the following are characteristics of both bacteria and fungi?

    (A) Cell wall, DNA, and plasma membrane
    (B) Nucleus, organelles, and unicellularity
    (C) Plasma membrane, multicellularity, and Golgi apparatus
    (D) Cell wall, unicellularity, and mitochondria
    (E) Nucleus, RNA, and cell wall

45. A sustained decrease in circulating Ca^{2+} levels might be caused by decreased levels of which of the following substances?

    (A) Growth hormone
    (B) Parathyroid hormone
    (C) Thyroid hormone
    (D) Calcitonin
    (E) Glucagon

46. The synthesis of new proteins necessary for lactose utilization by the bacterium *E. coli* using the *lac* operon is regulated

    (A) by the synthesis of additional ribosomes.
    (B) at the transcription stage.
    (C) at the translation stage.
    (D) by differential replication of the DNA that codes for lactose-utilizing mechanisms.
    (E) by a negative-feedback system.

47. Which of the following statements about enzymes is NOT true?

    (A) They are organic compounds made of proteins.
    (B) They are catalysts that alter the rate of a reaction.
    (C) They are operative over a wide pH range.
    (D) The rate of catalysis is affected by the concentration of substrate.
    (E) They denature if exposed to high temperatures.

48. A population of paramecium has the scientific name *P. aurelia* whereas another population has the name *P. caudatum*. This method of classification indicates that the organisms do NOT belong to the same

    (A) phylum
    (B) class
    (C) family
    (D) genus
    (E) species

49. In DNA replication, which of the following does NOT occur?

    (A) Helicase unwinds the double helix.
    (B) DNA ligase links the Okazaki fragments.
    (C) RNA polymerase is used to elongate both chains of the helix.
    (D) DNA strands grow in the 5´ to 3´ direction.
    (E) Complementary bases attach to each DNA strand.

50. A change in a neuron membrane potential from +50 millivolts to –70 millivolts is considered

    (A) depolarization
    (B) repolarization
    (C) hyperpolarization
    (D) an action potential
    (E) saltatory conduction

GO ON TO THE NEXT PAGE

51. The energy given up by electrons as they move through the electron transport chain is used to

(A) break down glucose
(B) make glucose
(C) produce ATP
(D) make NADH
(E) make $FADH_2$

52. If a photosynthesizing plant began to release $^{18}O_2$ instead of normal oxygen, one could most reasonably conclude that the plant had been supplied with

(A) H_2O containing radioactive oxygen
(B) CO_2 containing radioactive oxygen
(C) $C_6H_{12}O_6$ containing radioactive oxygen
(D) NO_2 containing radioactive oxygen
(E) oxygen from the atmosphere

53. All of the following statements describe the unique characteristics of water EXCEPT:

(A) It is a polar solvent.
(B) It is relatively resistant to temperature changes.
(C) It forms hydrogen bonds with disaccharides.
(D) It can dissociate into hydrogen ions and hydroxide ions.
(E) It is a hydrophobic solvent.

54. Chemical substances released by organisms that elicit a physiological or behavioral response in other members of the same species are known as

(A) auxins
(B) hormones
(C) pheromones
(D) enzymes
(E) coenzymes

55. Homologous structures are often cited as evidence for the process of natural selection. All of the following are examples of homologous structures EXCEPT

(A) the wings of a bird and the wings of a bat
(B) the flippers of a whale and the arms of a man
(C) the pectoral fins of a porpoise and the flippers of a seal
(D) the forelegs of an insect and the forelimbs of a dog
(E) the hind legs of a lizard and the legs of a chicken

56. The sliding action of actin and myosin in skeletal muscle contraction requires which of the following?

I. Ca^{2+}

II. ATP

III. actin

(A) I only
(B) II only
(C) I and III
(D) II and III
(E) I, II, and III

57. Certain populations of finches have long been isolated on the Galapagos islands off the western coast of South America. Compared with the larger stock population of mainland finches, these separate populations exhibit far greater variation over a wider range of species. The variation among these numerous finch species is the result of

(A) convergent evolution
(B) divergent evolution
(C) disruptive selection
(D) stabilizing selection
(E) directional selection

58. Which of the following contributes the MOST to genetic variability in a population?

(A) Mitosis
(B) Sporulation
(C) Binary fission
(D) Vegetative propagation
(E) Mutation

GO ON TO THE NEXT PAGE

Directions: Each group of questions consist of five lettered headings followed by a list of numbered phrases or sentences. For each numbered phrase or sentence select the one heading that is most closely related to it and fill in, the corresponding oval on the answer sheet. Each heading may be used once, more than once, or not at all in each group.

Questions 59–62

    (A) Analogous structures
    (B) Homologous structures
    (C) Vestigial structures
    (D) Convergent evolution
    (E) Divergent evolution

From the terms listed above, choose the one that most clearly accounts for the similarities between the members of each pair listed below.

59. The lung of a reptile and the air bladder of a fish

60. The spine of a sea urchin and the quill of a porcupine

61. The appendix of a human and the hipbone of a whale

62. The bearing of live young in marsupial moles and placental moles

Questions 63–65

    (A) Stomata
    (B) Lenticel
    (C) Palisade
    (D) Stroma
    (E) Cuticle

63. Porous area that facilitates gas exchange in woody stems

64. Tiny opening that regulates gas exchange in the leaf

65. Layer of leaf mesophyll with chloroplast-containing cells

GO ON TO THE NEXT PAGE

Questions 66–69

(A) Sporulation
(B) Budding
(C) Regeneration
(D) Binary fission
(E) Vegetative propagation

66. Asexual reproduction in bacteria

67. The restoration of severed appendages

68. Asexual reproduction in yeasts in which smaller cells grow from a parent cell

69. Asexual reproduction in which new plants develop from roots, stems, or leaves of the parent plant

Questions 70–73

(A) Nucleic acid
(B) Protein
(C) Cellulose
(D) Triglyceride
(E) Glycogen

70. Lipid that consists of three fatty acids covalently bonded to glycerol

71. The stored form of sugar in humans

72. A macromolecule that consists of a polymer of amino acids

73. A substance that cannot be broken down by cows

GO ON TO THE NEXT PAGE

Questions 74–77

(A)

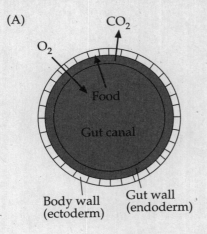

O$_2$

CO$_2$

Food

Gut canal

Body wall (ectoderm)   Gut wall (endoderm)

(B)

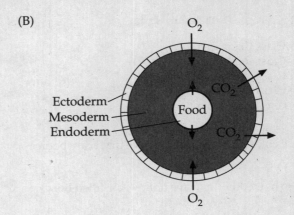

O$_2$

Ectoderm
Mesoderm
Endoderm

Food

CO$_2$

CO$_2$

O$_2$

(C)

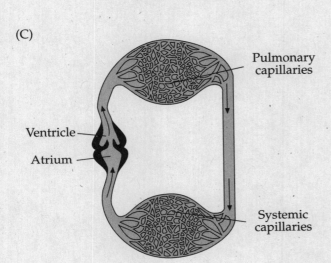

Pulmonary capillaries

Ventricle

Atrium

Systemic capillaries

(D)

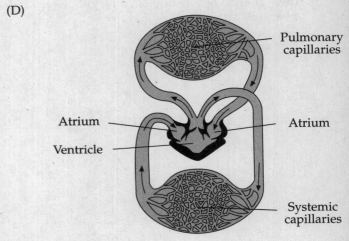

Pulmonary capillaries

Atrium

Ventricle

Atrium

Systemic capillaries

(E)

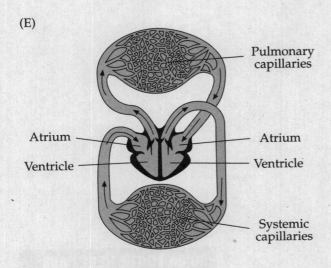

Pulmonary capillaries

Atrium

Ventricle

Atrium

Ventricle

Systemic capillaries

74. The circulatory system in birds

75. The circulatory system in hydras

76. The circulatory system in fish

77. The circulatory system in amphibians

GO ON TO THE NEXT PAGE

Questions 78–81

(A) Class Crustacea
(B) Class Reptilia
(C) Class Pisces
(D) Class Aves
(E) Class Amphibia

78. These cold-blooded animals have dry, scaly bodies

79. These homeotherms have hollow, air-filled bones

80. These cold-blooded animals lay eggs with leathery shells

81. These animals respire through their gills, lungs, and moist, soft skin

GO ON TO THE NEXT PAGE

Directions: Each group of questions below concerns an experimental or laboratory situation or data. In each case, first study the description of the situation or data. Then choose the one best answer to each question following it and fill in the corresponding oval on the answer sheet.

Questions 82–84 refer to the following information and graph.

A marine ecosystem was sampled in order to determine its food chain. The results of the study are shown below.

Type of Organism	Number of Organism
Shark	2
Small crustaceans	400
Mackerel	20
Plankton	1000
Herring	100

82. Which of the following organisms in this population are secondary consumers?

(A) Sharks
(B) Mackerels
(C) Herrings
(D) Shrimp
(E) Phytoplanktons

83. Which of the following organisms has the largest biomass in this food chain?

(A) Phytoplanktons
(B) Mackerels
(C) Herrings
(D) Shrimp
(E) Sharks

84. If the herring population is reduced by predation, which of the following is most likely to occur in this aquatic ecosystem?

(A) The mackerels will be the largest predator in the pond.
(B) The shrimp population will be greatly reduced.
(C) The plankton population will be reduced over the next year.
(D) The shrimp will become extinct.
(E) There will be no change in the number of sharks in the pond.

GO ON TO THE NEXT PAGE

Questions 85–87 refer to the following information and graph.

To understand the workings of neurons, an experiment was conducted to study the neural pathway of a reflex arc in frogs. A diagram of a reflex arc is given below.

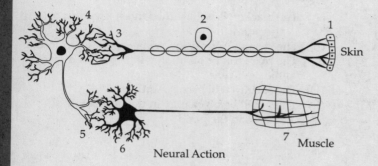

Neural Action

85. Which of the following represents the correct pathway taken by a nerve impulse as it travels from the spinal cord to effector cells?

(A) 1-2-3-4
(B) 6-5-4-3
(C) 2-3-4-5
(D) 4-5-6-7
(E) 7-6-5-4

86. The brain of the frog is destroyed. A piece of acid-soaked paper is applied to the frog's skin. Every time the piece of paper is placed on its skin, one leg moves upward. Which of the following conclusions is best supported by the experiment?

(A) Reflex actions are not automatic.
(B) Some reflex actions can be inhibited or facilitated.
(C) All behaviors in frogs are primarily reflex responses.
(D) This reflex action bypasses the brain.
(E) Reflex responses account for the total behavior in frogs.

87. A nerve impulse requires the release of neurotransmitters at the axonal bulb of a presynaptic neuron. Which of the following best explains the purpose of neurotransmitters, such as acetylcholine?

(A) They speed up the nerve conduction in a neuron.
(B) They open the sodium channels in the axonal membrane.
(C) They excite or inhibit the postsynaptic neuron.
(D) They open the potassium channels in the axonal membrane.
(E) They force potassium ions to move against the concentration gradient within the axonal membrane.

GO ON TO THE NEXT PAGE

Questions 88–90 refer to the following graph and information.

An experiment was conducted to observe the light-absorbing properties of chlorophylls and carotenoids using a spectrophotometer. The pigments were first extracted and dissolved in a solution. They were then illuminated with pure light of different wavelengths to detect which wavelengths were absorbed by the solution. The results are presented in the absorption spectrum below.

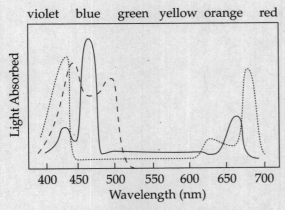

Absorption Spectrum for Green Plants

············ Chlorophyll *a*
———— Chlorophyll *b*
- - - - Carotenoids

88. At approximately what wavelength does chlorophyll *a* maximally absorb light?

(A) 425 nm
(B) 450 nm
(C) 575 nm
(D) 640 nm
(E) 680 nm

89. Which of the following color ranges were strongly absorbed by the chlorophyll pigments?

(A) Yellow-green and orange-red
(B) Violet-blue and orange-red
(C) Blue-green and yellow-orange
(D) Blue-green and orange-red
(E) Orange-red and yellow-green

90. Which of the following conclusions can be drawn from the experiment?

(A) Light reflected by the pigments is involved in photosynthesis.
(B) All wavelengths of light that reach the leaf can be utilized for photosynthesis.
(C) The light absorbed by the pigments drives photosynthesis.
(D) All wavelengths of visible light are absorbed and used by the plant.
(E) The light reflected from leaves consists mainly of violet-blue light.

GO ON TO THE NEXT PAGE

Questions 91–92 refer to the following graph. A pedigree was established to trace the colorblindness allele through four generations.

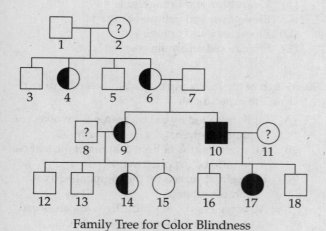

Family Tree for Color Blindness

91. Based on the pedigree above, what is the probability that a male child born to individuals 6 and 7 will be color-blind?

(A) 0

(B) $\frac{1}{4}$

(C) $\frac{1}{2}$

(D) $\frac{3}{4}$

(E) 1

92. In this family tree, individual 2 can best be classified as a

(A) normal male
(B) normal noncarrier female
(C) color-blind male
(D) carrier female
(E) color-blind female

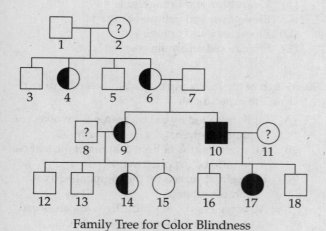

GO ON TO THE NEXT PAGE

Questions 93–95 refer to the figure and chart below.

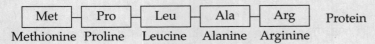
mRNA

| Met | Pro | Leu | Ala | Arg | Protein |
| Methionine | Proline | Leucine | Alanine | Arginine | |

Formation of a Protein

First Position (5' end)	Second Position	The Genetic Code: Condons of mRNA that Specify a Given Amino Acid			
		Third Position (3' end)			
		U	C	A	G
U	U	UUU UUC Phenylalanine		UUA UUG Lucine	
	C	UCU UCC UCA UCG Serine			
	A	UAU UAC Tytrosine		UAA UAG	
	G	UGU UGC Cysteine		UGA	UGG Tryptophan
C	U	CUU CUC CUA CUG Leucine			
	C	CCU CCC CCA CCG Proline			
	A	CAU CAC Histidine		CAA CAG Glutamine	
	G	CGU CGC CGA CGG Arginine			
A	U	AUU AUC AUA Isoleucine			AUG
	C	ACU ACC ACA ACG Threonine			
	A	AAU AAC Asparagine		AAA AAG Lysine	
	G	AGU AGC Serine		AGA AGG Arginine	
G	U	GUU GUC GUA GUG Valine			
	C	GCU GCC GCA GCG			
	A	GAU GAC Aspartic Acid		GAA GAG Glutamic acid	
	G	GGU GGC GGA GGG Glycine			

93. Which of the following DNA strands would serve as a template for the amino acid sequence shown above?
(A) 3´-ATGCGACCAGCACGT- 5´
(B) 3´-AUGCCACUAGCACGU- 5´
(C) 3´-TACGGTGATCGTGCA- 5´
(D) 3´-UACGGUGAUCGUGCA- 5´
(E) 3´-TGCACGATCACCGTA- 5´

94. If a mutation occurs in which uracil is deleted from the messenger RNA after methionine is translated, which of the following represents the resulting amino acid sequence?
(A) serine-histidine-serine-threonine
(B) methionine-proline-glutamine-histidine
(C) methionine-proline-leucine-alanine-arginine
(D) methionine-proline-alanine-arginine-arginine
(E) serine-proline-leucine-alanine-arginine

95. The mRNA above was found to be much smaller than the original mRNA synthesized in the nucleus. This is due to the
(A) addition of a poly(A) tail to the mRNA molecule
(B) addition of a cap to the mRNA molecule
(C) excision of exons from the mRNA molecule
(D) excision of introns from the mRNA molecule
(E) translation of the mRNA molecule

GO ON TO THE NEXT PAGE

Questions 96–98 refer to the following chart.

The loss of water by evaporation from the leaf openings is known as transpiration. The transpiration rates of various plants are shown below.

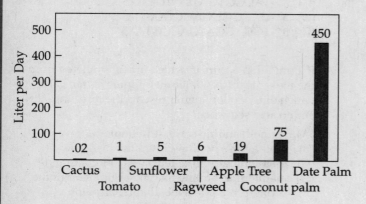

Transpiration Rates for Plants

96. How many liters of water per week are lost by a coconut palm?

(A)   19
(B)   25
(C)   75
(D)   450
(E)   525

97. During transpiration, water passes from the soil to the leaves via

(A)   tracheids
(B)   sieve-tube cells
(C)   lenticels
(D)   companion cells
(E)   spongy cells

98. Transpiration aids in the transport of water by doing all of the following EXCEPT

(A)   directing the upward movement of water to leaves
(B)   keeping the air spaces of leaves moist
(C)   increasing water pressure in the roots
(D)   contributing to the cooling of the plant through exudation
(E)   increasing the amount of food absorbed by the plant

GO ON TO THE NEXT PAGE

Questions 99–100 refer to the following information.

A scientist studies the storage and distribution of oxygen in humans and Weddell seals to examine the physiological adaptations that permit seals to descend to great depths and stay submerged for extended periods. The figure below depicts the oxygen storage in both organisms.

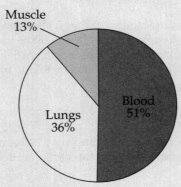

Human (70 kilograms)

Total oxygen store: 1.95 liters

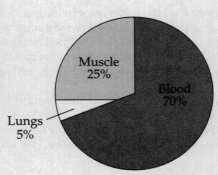

Weddell seal (450 kilograms)

Total oxygen store: 25.9 liters

99. Compared with humans, approximately how many liters of oxygen does the Weddell seal store per kilogram of body weight?

(A) The same amount of oxygen
(B) Twice the amount of oxygen
(C) Three times the amount of oxygen
(D) Five times the amount of oxygen
(E) Thirteen times the amount of oxygen

100. During a dive, a Weddell seal's blood flow to the abdominal organs is shut off and oxygen-rich blood is diverted to the eyes, brain, and spinal cord. Which of the following is the most likely reason for this adaptation?

(A) To increase the number of red blood cells in the nervous system
(B) To increase the amount of oxygen reaching the skeletomuscular system
(C) To increase the amount of oxygen reaching the central nervous system
(D) To increase the oxygen concentration in the lungs
(E) To decrease the extreme pressure on the diving seal

STOP

IF YOU FINISH BEFORE TIME IS CALLED, YOU MAY
CHECK YOUR WORK ON THIS SECTION.

DO NOT GO ON UNTIL YOU ARE TOLD TO DO SO.

BIOLOGY
SECTION II

Planning time—10 minutes

Writing time—1 hour and 30 minutes

You will have 10 minutes to read the exam questions. Spend this time reading through all of the questions, noting possible problem-solving approaches and otherwise planning your answers. It's fine to make notes on the green question insert, but be sure to write your answers and anything else that might be worth partial credit in the pink answer booklet—the graders will not see the green insert. After 10 minutes you will be told to break the seal on the pink Free-Response booklet and begin writing your answers in that booklet.

Answer all questions. Number your answer as the question is numbered below.

Answers must be in essay form. Outline form is NOT acceptable. Labeled diagrams may be used to supplement discussion, but in no case will a diagram alone suffice. It is important that you read each question completely before you begin to write.

1. The cell membrane is an important structural feature of a nerve cell.
   a. Describe how the cell membrane of a neuron is similar to the cell membrane of other cells in the transport of materials across a membrane.
   b. Discuss what ions are associated with the transmission of a nerve impulse.
   c. Describe the role of membranes in the conduction of a nerve impulse.

2. Before plants could survive on land several adaptational problems had to be solved. Describe the problems associated with plant survival on land and discuss three structural adaptations that kept plants supplied with water. Describe **two** other structures that contributed to their success in a terrestrial environment.

3. Sickle cell anemia is a genetic disorder caused by the abnormal gene for hemoglobin S. A single substitution occurs in which glutamic acid is substituted for valine in the sixth position of the hemoglobin molecule. This change reduces hemoglobin's ability to carry oxygen.
   a. Discuss the type of mutation that occurs in base substitution.
   b. Biologists used gel electrophoresis to initially identify the mutant gene. Explain how gel electrophoresis could be applied to the identification of the gene mutation. Discuss the use of restriction enzymes.
   c. Hemoglobin S is transmitted as a simple Mendelian allele. Describe the outcome if a female who does not carry the abnormal allele mates with a male homozygous for the disease. If any of the children are heterozygotes they are considered carriers.

4. Discuss the Krebs cycle, the electron transport chain, and oxidative phosphorylation.
   a. Explain why these steps are considered aerobic processes.
   b. Discuss the location at which <u>each</u> stage occurs.
   c. Discuss the role of NADH and $FADH_2$ in aerobic respiration.

END OF EXAMINATION

18

Answers and Explanations to Practice Test 1

1. **(E) is the correct answer.** The resting potential depends on active transport (the Na^+/K^+ pump) and the selective permeability of the axon membrane to K^+ than to Na^+, which leads to a differential distribution of ions across the axonal membrane.

2. **(B) is the correct answer.** The Krebs cycle occurs in the mitochondrial matrix. Don't forget to review the site of each stage of aerobic respiration. (A): Glycolysis, the first step in aerobic respiration, occurs in the cytoplasm. (C): The electron transport chain occurs along the inner mitochondrial membrane. (E): Oxidative phosphorylation occurs as protons (H^+ ions) move from the intermembrane space to the mitochondrial matrix.

3. **(A) is the correct answer.** The earth's major land biomes are classified according to the climate zones and organisms in a particular region. An individual traveling up a mountain will probably encounter four biomes. Tropical regions (tropical rain forests) are known for their abundant rain supply and high temperatures. Temperate deciduous forests are similar to regions in North America with cold winters, dry summers, and trees that shed leaves. This is followed by the taiga biome. Taiga includes the evergreen conifer trees and animals such as moose and black bears. The top of the mountain is a cold, treeless region with little vegetation—the tundra.

4. **(B) is the correct answer.** In meiosis, the sister chromatids separate during the second metaphase of meiosis (Meiosis II) whereas the sister chromatids separate during metaphase of mitosis. (A): In meiosis, there are two rounds of cell division whereas in mitosis, there is only one round of cell division. (C): Chromosomes are replicated during interphase in both meiosis and mitosis. (D): Spindle fibers form during prophase in both mitosis and meiosis. (E): In meiosis, there is a reduction in the number of chromosomes whereas in mitosis, the number of chromosomes remains the same.

5. **(C) is the correct answer.** (A), (B), and (D): Amino acids are organic molecules that contain carbon, hydrogen, oxygen, and nitrogen. Don't forget to associate amino acids with nitrogen because of the amino group (NH_2). (E): Amino acids do not contain phosphorus.

6. **(B) is the correct answer.** Unlike eukaryotes, prokaryotes (which include bacteria) do not contain membrane-bound organelles. (A), (D), and (E): Bacteria contain circular double-stranded DNA, ribosomes, and a cell wall. (C): Bacterial cell membranes are made up of a bilipid layer with proteins interspersed.

7. **(D) is the correct answer.** Populations, not individuals, evolve over time. That's why the population is the functional unit of evolution. (A): Genetic changes occur only at the individual level. (B): Only under Hardy-Weinberg equilibrium does the gene pool remain fixed over time in a population. However, this statement does not explain why the population is the evolving unit. (C): This statement is true but does not address the question. (E): This statement is also true but it does not address the question.

8. **(D) is the correct answer.** The medulla controls basic involuntary actions such as breathing. Don't forget to review the other parts of the brain. (A): The cerebellum controls and coordinates movement and balance. (B): The cerebrum controls higher-level thinking and reasoning. (C): The hypothalamus maintains homeostasis. (E): The thalamus integrates information from other parts of the brain. We did not discuss the thalamus. This was a great opportunity to use Process of Elimination.

9. **(C) is the correct answer.** The vascular cambium includes actively dividing cells that lie between the xylem and phloem to produce secondary xylem and phloem. (A): The apical meristem refers to actively dividing cells located at the tip of roots and stems. (B): The epidermis is the outermost top layer in leaves. (D): Cork cambium is the outer layer of the meristem in woody stems. (E): Lenticels are the pores used for gas exchange in woody stems.

10. **(B) is the correct answer.** In order to determine the genotype of the parents, use the ratio of the offspring given in the question and work backward. The ratio of black-haired to white-haired guinea pigs is 3:1. The offspring were therefore BB, Bb, Bb, and bb. Now draw a Punnett square to figure out the genotype of the parents. The parents are Bb and Bb.

11. **(D) is the correct answer.** The mean weight of the offspring in the next generation will be heavier than the mean weight of the original population because all the lighter horses in the original population died off. The normal distribution for weight will therefore shift to the heavier end (to the right of the graph). You can therefore eliminate answer choices (C) and (E) because the mean weight should increase. (A) and (B): The mean weight of the offspring could be heavier or lighter than their parents.

12. **(D) is the correct answer.** A fertilized egg undergoes a series of changes in the following order: cleavage (first divisions of the zygote), morula (solid ball), blastula (hollow ball), gastrula (formation of three germ layers), and neurula (formation of the nervous system).

13. **(A) is the correct answer.** The primitive atmosphere lacked oxygen (O_2). It contained methane (CH_4), ammonia (NH_3), hydrogen (H_2), and water (H_2O).

14. **(C) is the correct answer.** Make a Punnett square to determine the probability that the couple has a child with blood type AB. The probability is $\frac{1}{4}$ whether it's the first child or the third child.

15. **(B) is the correct answer.** You're looking for the best explanation for the low trophic level efficiency among elks. The best explanation is that large herbivores, such as elks, eat plants but only incorporate a small portion of the nutrients in plants because they cannot digest cellulose. (A): This answer choice sounds good but we don't know if elks consume less food than do other herbivores. (C): This response also sounds good but we don't know if elks use a lot more energy than do other herbivores as they search for food. (D): This response is clearly false. Eating producers is a very efficient way of getting energy. (E): We don't know if elks lose more body heat than do other herbivores. Even if they did, this is not directly related to trophic level efficiently.

16. **(A) is the correct answer.** Porifera sponges and coelenterates are more primitive than platyhelminthes (such as flatworms), so cross out (D) and (E). Echinoderms (such as starfish) and chordates are more complex than arthropods, so cross out (B) and (C).

17. **(B) is the correct answer.** Pepsin works in the stomach (not the small intestine) to break down proteins to peptides. Complete digestion occurs in the small intestine. (A): Pancreatic lipase breaks down fats into three fatty acids and glycerol. (C): Pancreatic amylase breaks down carbohydrates into simple sugars. (D): Bile emulsifies lipids and makes them more accessible to lipase. (E): Food is absorbed by the villi in the small intestine.

18. **(E) is the correct answer.** Ribosomes are the site of protein synthesis. Therefore, the correct answer should start with ribosome. So eliminate answer choices (A), (B), and (D). The polypeptide then moves through the rough ER to the Golgi apparatus where it is modified and packaged into a vesicle. The vesicle then floats to the plasma membrane and is secreted.

19. **(E) is the correct answer.** This statement is false because an individual with two identical alleles is said to be *homozygous* not heterozygous with respect to that gene. (A) and (B): Alleles are different forms of the same gene found on corresponding positions of homologous chromosomes. (C): More than two alleles can exist for a gene but a person can have only two alleles for each trait. (Remember our discussion of blood groups in Chapter 11.) (D): One allele can mask another allele.

20. **(C) is the correct answer.** To make multiple copies of a plasmid (a small circular DNA) it should be inserted into a bacterium. (A): A plasmid would not replicate if it were inserted into a virus. (B): If a plasmid were treated with a restriction enzyme it would be cut into smaller fragments. This would not give us cloned versions of the plasmid. (D): If the plasmid were run on a gel (using gel electrophoresis) this would only tell us the size of the plasmid. (E): Plasmids are small circular DNA and cannot be infected. Only living organisms can be infected.

21. **(D) is the correct answer.** The least likely explanation for why mutations are low is that mutations produce variability in a gene pool. Any gene is bound to mutate. This produces a constant input of new genetic information in a gene pool. This answer choice doesn't give us any additional information about the rate of mutations. (A): Some mutations are subtle and cause only a slight decrease in reproductive output. (B): Some mutations are harmful and decrease the productive success of the individual. (C): Some mutations are deleterious and lead to total reproductive failure. The zygote fails to develop. (E): Some mutations back mutate. By the way, this is a tough question and should have been skipped on both a first *and* second pass!

22. **(C) is the correct answer.** Desert plants (for example, cacti) have all the features of flowering plants (vascular tissues, multiple leaves, megaspores, and stomates). An adaptive feature in this type of plant is thick cuticles to prevent excessive water loss.

23. **(B) is the correct answer.** Legume plants are able to live in nitrogen-poor soil because they obtain nitrogen from nitrogen-fixing bacteria. (A), (C), and (D): These plants cannot make their own proteins without nitrogen from nitrogen-fixing bacteria. (E): Legumes do not possess nitrogen-absorbing root hairs.

24. **(E) is the correct answer.** Don't forget to review the organs that arise from the three primary germ layers. (A), (B), (C), and (D): The brain, eyes, skin, and nervous system are derived from the ectoderm. (E): Ovaries, which are part of the reproductive system, are derived from the mesoderm.

25. **(D) is the correct answer.** Deoxygenated blood from the vena cava enters the right atrium (2), then the right ventricle (1), and then enters the pulmonary arteries (4). The left atrium (5), left ventricle (6), and aorta (3) all carry oxygenated blood.

26. **(E) is the correct answer.** The egg yolk provides food for the embryo. (A): The amnion protects the embryo. (B): The chorion is the outermost layer that surrounds the embryo. (C): The placenta provides nourishment for the embryo in mammals, not in birds. (D): The ovary is the reproductive organ in which eggs mature.

27. **(B) is the correct answer.** Normal cells can become cancerous when a virus invades the cell and takes over the replicative machinery. (A): A pili forms between two bacteria. (C): The host's genome is not *converted* to the viral genome. (D): Spores are released by fungi not viruses. (E): Incorporation of free-floating DNA would not necessarily produce cancerous cells.

28. **(C) is the correct answer.** The primary difference between bone and cartilage is that cartilage is flexible and lacks blood vessels. (A): They are both part of the skeletal system. (B) and (E): Bone and cartilage are connective tissues made up of collagen and calcium salts. (D): Both cartilage and bones secrete a matrix.

29. **(D) is the correct answer.** If potential mates have similar breeding seasons they will most likely mate. Answer choices (A), (B), (C), and (E) are all examples of prezygotic barriers that can prevent interbreeding. Use common sense to eliminate the other answer choices. (A): This is an example of ecological isolation. If the organisms don't meet they won't reproduce. (B) and (C): If the potential mates do not share the same behaviors (such as courtship rituals) they may not mate. (E) This is an example of physiological isolation. If the gametes are biochemically different the organisms won't reproduce.

30. **(D) is the correct answer.** There is no union of gametes in mitosis. (A) and (C): Asexual reproduction involves the production of two new cells with the same number of chromosomes as the parent cell. If the parent cell is diploid, then the daughter cells will be diploid. (B): The daughter cells are identical to the parent cell. (E): Chromosomes replicate during interphase.

31. **(D) is the correct answer.** Arteries are thick-walled vessels that carry blood *away* from the heart. Blood moves by contracting muscles. (A), (B), (C), and (E) are all characteristics of veins. Veins are thin-walled vessels (with valves) that return blood to the heart.

32. **(B) is the correct answer.** Auxins are plant hormones with many properties. They stimulate cell elongation, bud and fruit development, and play a role in phototropism. (A): Cytokinins stimulate cell division in plants. (C): Gibberellins promote stem elongation (especially in dwarf plants). (D): Abscisic acid causes growth inhibition. (E): Ethylene causes fruit ripening.

33. **(A) is the correct answer.** After double fertilization, the ovule becomes the seed. (B) and (E): The ovary becomes the fruit and the fertilized egg becomes the embryonic plant. (D): The cotyledons are the leaves of the embryonic plant. (C): The pollen grains form the male gametes.

34. **(E) is the correct answer.** The frequency of the homozygous dominant genotype (AA) is 0.49. To find the dominant *allele* frequency, we can use the formula provided by the Hardy-Weinberg theory: $p^2 + 2pq + p^2 = 1$, where p represents the dominant allele and q represents the recessive allele. Since we know that p^2 represents the frequency of the homozygous dominant genotype, we can find the frequency of the dominant allele (p) by taking the square root of the frequency of the genotype. The square root of 0.49 is 0.7. Note that by using the formula $p + q = 1$, we can also determine the frequency of the recessive allele (q). It would be 0.3 ($0.7 + q = 1$).

35. **(A) is the correct answer.** Crossing over (exchange of genetic material) occurs in prophase I of meiosis. (B): During metaphase I the tetrads line up at the metaphase plate. (C): During anaphase, the tetrads separate. (D): During prophase II, the chromosomes split at the centromere. (E): During metaphase II, the chromosomes line up at the metaphase plate.

36. **(D) is the correct answer.** The microspores (male gametes) are produced in the pollen grains of the anther not the stamens. All of the other statements are true. (A): Double fertilization occurs in the ovules. (B): A megaspore divides and becomes the female gametes. (C): The pollen tube grows down the style to the ovary. (E): The endosperm (3n) becomes the food for the embryo.

37. **(E) is the correct answer.** Bryophytes are primitive plants that include mosses and liverworts. (A): They do not have a protective layer around their gametes. The gametes are found in the parent gametophyte. (B): They lack conductive tissues. (C) and (D): They lack true leaves. They therefore do not have stomates or waxy cuticles.

38. **(E) is the correct answer.** The net primary productivity is the energy producers have left for storage after their own energy needs have been met. Energy stored = (Total energy produced)–(energy used in cellular respiration). (A): It is not the energy available to producers. (B): It is not the total chemical energy in producers. That's the gross primary productivity. (C): It is not the biomass (total living material) among producers. (D): It is not the energy used in respiration by heterotrophs. Primary productivity always refers to producers.

39. **(C) is the correct answer.** (A), (B), (E): These organisms exhibit the same behavior because they were subjected to the same environmental conditions and similar habitats. This is an example of convergent evolution. (C): However, they are not genetically similar. (One is an insect and the other a bird.) (D): They are analogous. They exhibit the same function but are structurally different.

40. **(C) is the correct answer.** These characteristics are all found in insects, which include arthropods with wings. (A): A snail is a mollusk. (B): A sponge is a porifera. (D): An earthworm is an annelid. (E): A flatworm is a platyhelminthes.

41. **(D) is the correct answer.** Beta cells secrete insulin. Destruction of beta cells in the pancreas will halt the production of insulin. Therefore, eliminate answer choices (A), (B) and (C). This will lead to an increase in blood glucose levels.

42. **(D) is the correct answer.** The sympathetic division is active during emergency situations. This leads to a decrease in peristalsis in your gastrointestinal tract. (Your stomach shuts down.) Stimulation of the sympathetic nervous system leads to (A): pupils dilating, (B): peripheral blood vessels constricting, (C): sweating, and (E): heart rate increasing.

43. **(C) is the correct answer.** This is an example of mutualism. Both organisms benefit. (A): Parasitism is an example of a symbiotic relationship in which one organism benefits and the other is harmed. (B): Commensalism is when one organism benefits and the other is unaffected. (D): Endosymbiosis is the idea that some organelles originated as symbiotic prokaryotes that live inside larger cells. (E): This is a behavior that is directly beneficial to other members of the species but at some risk to the organism.

44. **(A) is the correct answer.** They both contain genetic material (DNA), a plasma membrane, and a cell wall. Use Process of Elimination. Unlike fungi, bacteria lack a definite nucleus. Therefore, eliminate (B) and (E). Bacteria are unicellular whereas fungi are both unicellular and multicellular. Therefore, eliminate (C) and (D).

45. **(B) is the correct answer.** Let's use Process of Elimination. The two hormones that are responsible for the maintenance of Ca^{2+} in the blood are parathyroid hormone and calcitonin. Therefore, eliminate answer choices (A), (C) and (E). Now review the effects of these hormones. Parathyroid hormone increases Ca^{2+} in the blood while calcitonin decreases Ca^{2+} in the blood. A sustained decrease in circulating Ca^{2+} levels might be caused by decreased levels of parathyroid hormone.

46. **(B) is the correct answer.** This question tests your understanding of what stage is responsible for the synthesis of new proteins for lactose utilization. The region of bacterial DNA that controls gene expression is the *lac* operon. Structural genes will be transcribed to produce enzymes, which produce an mRNA involved in digesting lactose.

47. **(C) is the correct answer.** (A) and (B): Enzymes, which are proteins, are organic catalysts that speed up reactions without altering them. They are not consumed in the process. (D): The rate of reaction can be affected by the concentration of the substrate up to a point. (E): They denature (uncoil) when they're exposed to high temperatures.

48. **(E) is the correct answer.** The binomial classification system is based on a genus and species name. (Don't forget the classification mnemonic). Therefore, the paramecia share the same genus (same first name) but different species (different second name).

49. **(C) is the correct answer.** DNA polymerase not RNA polymerase is the enzyme that causes the DNA strands to elongate. (A): DNA helicase unwinds the double helix. (B): DNA ligase seals the discontinuous Okazaki fragments. (D) and (E): DNA strands, in the presence of DNA polymerase, always grow in the 5' to 3' direction as complementary bases attach.

50. **(B) is the correct answer.** A voltage change from +50 to –70 is called repolarization. (A): A voltage change from –70 to +50 is called depolarization. (C): A voltage change from –70 to –90 is called hyperpolarization. (D): An action potential is a traveling depolarized wave. It refers to the whole thing: from depolarization, repolarization, hyperpolarization, and back to a resting potential. (E): Saltatory conduction refers to the skipping of an impulse from one node to another node in myelinated neurons.

51 **(C) is the correct answer.** Electrons passed down along the electron transport chain from one carrier to another lose energy and provide energy for making ATP. (A): Glucose is decomposed during glycolysis but this process is not associated with energy given up by electrons. (B): Glucose is made during photosynthesis. (D) and (E): NADH and $FADH_2$ are energy-rich molecules, which accepted electrons during the Krebs cycle.

52. **(A) is the correct answer.** The oxygen released during the light reaction comes from the splitting of water. (Review the reaction for photosynthesis.) Therefore water must have originally contained the radioactive oxygen. (B): Carbon dioxide is involved in the dark reaction and produces glucose. (C): Glucose is the final product and would not be radioactive unless carbon dioxide was the radioactive material. (D): Nitrogen is not part of photosynthesis. (E): Atmospheric oxygen has nothing to do with photosynthesis.

53. **(E) is the correct answer.** Hydrophobic means "water fearing;" so water cannot be a hydrophobic solvent (A) and (C): Water is a polar solvent; it contains both negatively and positively charged ends that can form hydrogen bonds with other polar substances. (B): It can resist temperature changes because it has a high specific heat. (You need lots of heat to raise the temperature of water.) (D): Water can break up into H^+ and OH^- ions.

54. **(C) is the correct answer.** Pheromones act as sex attractants, alarm signals, or territorial markers. Use Process of Elimination for this question. (A): Auxins are plant hormones that promote growth. (B): Hormones are chemical messengers that produce a specific effect on target cells within the same organism. (D): Enzymes are catalysts that speed up reactions. (E): Coenzymes are organic substances that assist enzymes in a chemical reaction.

55. **(D) is the correct answer.** Homologous structures are organisms with the same structure but different functions. The forelegs of an insect and the forelimbs of a dog are not structurally similar. (One is an invertebrate and the other a vertebrate.) They do not share a common ancestor. However, both structures are used for movement. All of the other examples are vertebrates which are structurally similar.

56. **(E) is the correct answer.** (I): Muscle contractions require calcium ions. (II): In order to have a muscle contraction you need energy—ATP. (III): Actin is one of the proteins involved in muscle contractions.

57. **(B) is the correct answer.** Speciation occurred in the Galapagos finches as a result of the different environments on the islands. This is an example of divergent evolution. The finches were geographically isolated. (A): Convergent evolution is the evolution of similar structures in distantly related organisms. (C): Disruptive selection is selection that favors both extremes at the expense of the intermediates in a population. (D): Stabilizing selection is selection that favors the intermediates at the expense of the extreme phenotypes in a population. (E): Directional selection is selection that favors one of the extremes in a population.

58. **(E) is the correct answer.** Mutations produce genetic variability. All of the other answer choices are forms of asexual reproduction.

59. **(B) is the correct answer.** These structures are homologous. They are structurally similar but have different functions. The swim (air) bladder is a gas-filled air sac that provides buoyancy for fish. It is homologous with the lungs of vertebrates.

60. **(A) is the correct answer.** These structures are analogous. They have the same function but they're different structures. Spines and quills are both used for protection.

61. **(C) is the correct answer.** These structures are vestigial. They are parts of an animal that are no longer functional.

62. **(D) is the correct answer.** This is an example of convergent evolution. These unrelated organisms (marsupial moles and placental moles) resemble each other—they bear live young. Convergent evolution is the independent evolution of similar structures in distantly related organisms that occur in similar ecological niches.

63. **(B) is the correct answer.** Lenticels are pores in stems that facilitate gas exchange.

64. **(A) is the correct answer.** Stomates are pores in the epidermis of leaves which allow the diffusion of gases.

65. **(C) is the correct answer.** Don't be thrown off by the term mesophyll. You know that the palisade layer contains photosynthetic cells.

66. **(D) is the correct answer.** Binary fission is the equal division of a bacterial cell into two. It is a form of asexual reproduction.

67. **(C) is the correct answer.** Regeneration is the growing back of a lost body part. Use Process of Elimination.

68. **(B) is the correct answer.** Budding is a form of asexual reproduction in yeasts whereby smaller cells grow from a parent cell.

69. **(E) is the correct answer.** Vegetative propagation is a form of asexual reproduction in which flowering plants can reproduce with the use of stems, leaves, or roots.

70. **(D) is the correct answer.** A triglyceride is an organic molecule that consists of three fatty acids and glycerol.

71. **(E) is the correct answer.** Glycogen is an animal starch. It is stored in the liver and muscles of animals.

72. **(B) is the correct answer.** A protein is a macromolecule consisting of polypeptides (amino acids).

73. **(C) is the correct answer.** Cellulose is a carbohydrate that can not be broken down by cows. That's why they need bacteria in their stomachs.

74. **(E) is the correct answer.** A bird has a four-chambered heart. Review the classification list in this book.

75. **(A) is the correct answer.** A hydra is a simple animal with two layers: an ectoderm and an endoderm. You can therefore eliminate (B). Review the classification list in this book.

76. **(C) is the correct answer.** A fish has a two-chambered heart. Review the classification list in this book.

77. **(D) is the correct answer.** An amphibian has a three-chambered heart. Review the classification list in this book.

78. **(B) is the correct answer.** Reptiles are usually covered with dry scales. Review the classification list in this book.

79. **(D) is the correct answer.** Homeotherms means warm-blooded. But even if you didn't know that, the only animals with air-filled bones are birds. Review the classification list in this book.

80. **(B) is the correct answer.** These are the characteristics of reptiles. Don't forget you can use an answer choice more than once. The egg is surrounded by a tough leathery shell, which protects the embryo against desiccation. Review the classification list in this book.

81. **(E) is the correct answer.** Amphibians have thin, moist skin. They were the first animals to conquer land so they have both gills and lungs. Review the classification list in this book.

82. **(C) is the correct answer.** Secondary consumers feed on primary consumers. If you set up a pyramid of numbers you'll see that the herrings belong to the third trophic level.

83. **(A) is the correct answer.** The biomass is the total bulk of a particular living organism. The plankton population has both the largest biomass and the most energy.

84. **(C) is the correct answer.** If the herring population decreases this will lead to an increase in the number of crustaceans and a decrease in the plankton population. Reorder the organisms according to their trophic levels and determine which populations will increase and decrease accordingly.

85. **(D) is the correct answer.** This question tests your ability to trace the neural pathway of a motor (effector) neuron. The nerve conduction will travel from the spinal cord (where interneurons are located) to the muscle.

86. **(D) is the correct answer.** Since the brain is destroyed it is not associated with the movement of the leg. (A): Reflex actions are automatic. (B) and (D): Both of these statements are true but are not supported by the experiment. (E): We do not have enough information from the passage to determine if this statement is true.

87. **(C) is the correct answer.** Neurotransmitters are released from the axonal bulb of one neuron and diffuse across a synapse to activate a second neuron. The second neuron is called a post-synaptic neuron. A neurotransmitter can either excite or inhibit the postsynaptic neuron. (A): The myelin sheath speeds up the conduction in a neuron. (B), (D), and (E): Both sodium and potassium channels open during an action potential. Neurotransmitters are not involved in actions related to the axon membrane. They do not force potassium ions to move against a concentration gradient.

88. **(A) is the correct answer.** If you look at the absorption spectrum you'll see that chlorophyll a has two peaks: one at 425 nm and 680 nm. Chlorophyll *a* maximally absorbs light at approximately 425 nm.

89. **(B) is the correct answer.** Chlorophyll pigments absorb light in the violet-blue and orange-red range and reflect green light. That's why plants appear green. Therefore you can eliminate answer choices that contain the color green. That's all of them except (B).

90. **(C) is the correct answer.** (A): Based on the results of the absorption spectrum, the light absorbed by chlorophylls and carotenoids is involved in photosynthesis. (B) and (D): We know from our review of photosynthesis that only certain wavelengths of visible light can be used for photosynthesis. (E): The light that is reflected is primarily green.

91. **(C) is the correct answer.** Draw a Punnett square for the couple and determine the probability of color-blindness for the boys. Individuals 6 and 7, (X^cX and XY) will produce two males: one X^cY and one XY. The probability of color-blindness is therefore 1/2.

92. **(D) is the correct answer.** Since the genotype of the mother is unknown we need to look at her children in order to determine her status. Since one of the sons is color-blind, she must have the color-blind gene. However, she is not color-blind because she produced a normal male.

93. **(C) is the correct answer.** The DNA template strand is complementary to the mRNA strand. Using the mRNA strand, work backward to establish the sequence of the DNA strand. Don't forget that DNA strands do not contain uracil so eliminate answer choices (B) and (D).

94. **(B) is the correct answer.** Use the amino acid chart to determine the sequence after uracil is deleted. The deletion of uracil creates a frameshift.

95. **(D) is the correct answer.** The mRNA is modified before it leaves the nucleus. It becomes smaller when introns (intervening sequences) are removed. (A) and (B): A poly (A) tail and a cap are added to the mRNA and would therefore increase the length of the mRNA. (C): Exons are the coding sequences that are kept by the mRNA. (E): Translation is when we produce a protein from a molecule of mRNA.

96. **(E) is the correct answer.** Make sure you read the question carefully. You are asked to calculate the number of liters per week, not per day. The chart tells us that a coconut palm loses 75 liters a day, which would mean 525 liters a week (7 x 75 = 525).

97. **(A) is the correct answer.** Water moves from the soil to the leaves via tracheids. (B) and (D): Sieve-tube cells and companion cells carry nutrients not water. (C): Lenticels are pores that regulate gas exchange in woody stems. (E): The spongy layer consists of loosely organized cells in the lower layer of the leaf.

98. **(E) is the correct answer.** Answer choices (A) through (D) all provide the force that raises water and solutes from the root to the stem. (E): Nutrients are carried through phloem and are not involved in transpiration.

99. **(B) is the correct answer.** The Weddell seal stores twice as much oxygen as humans. Calculate the liters per kilograms weight for both the seal and man using the information at the bottom of the chart. The Weddell seal stores 0.058 liters/kilograms (25.9 liters/450 kilograms) compared to 0.028 liters/kilograms (1.95 liters/70 kilograms) in humans.

100. **(C) is the correct answer.** The most plausible answer is that blood is redirected toward the central nervous system, which permits the seal to navigate for long durations. (A): The seal does not need to increase the number of red blood cells in the nervous system. (B): The seal does not need to increase the amount of oxygen to the skeletal system. (D): The diversion of blood does not increase the concentration of oxygen in the lungs. (E): The seal does not decrease pressure on itself if it diverts blood to the eyes, brain, and spinal cord.

SECTION II: FREE-RESPONSE QUESTIONS

In the following pages you'll find two types of aids for grading your essays: checklists and written paragraphs. You'll recall from our discussion of the essays in Chapter 14 that ETS uses checklists just like these to grade your essays on the actual test. They're actually quite simple to use.

For each item you mentioned, give yourself the appropriate number of points (1 point, 2 points, etc.). However, don't be kind to yourself just because you like your own essay: If the checklist mentions an explanation of structure and function and you failed to give both, then do not give yourself a point. ETS is very particular about this. You need to mention precisely the things they've listed, in the way they've listed them, in order to gain points. What if something you've mentioned doesn't appear on the list?

Provided you know it's valid, give yourself a point. If it was something you pulled out of your hat at the last minute, odds are it's not directly applicable to the question. However, since you might come up with details even more specific than those contained in the checklist, it's not unlikely that the example or structure you've cited is perfectly valid. If so, go ahead and give yourself the point. Remember that ETS hires college professors and high school teachers to read your essays, so they'll undoubtedly recognize any legitimate information you slip into the essay.

The second part of the answer key involves short paragraphs. These are not templates: ETS does not expect you to write this way. They are simply additional tools to help illustrate the things you need to squeeze into your essay in order to rack up the points. They explain in some detail how the various parts of the checklist relate to one another, and may give you an idea about how best to integrate them into your own essays come test-time.

If you find that you didn't do too well on the essay portion, go back and practice some essays from Chapter 14. You can use the chapters themselves as checklists. If you find it too difficult to grade your own essays, see if your teacher or a classmate will help you out. Good luck!

ESSAY CHECKLIST QUESTION 1

a. **Cellular structures**—4 points maximum

Neuronal plasma membranes and other cell membranes have—2 points maximum
  Bilipid layer (enriched with phospholipids and proteins)
  Selective permeability (regulates the passage of certain ions across its membrane)
Other structures—2 points maximum
  Nucleus (controls the activities of the cell)
  Ribosomes (synthesizes proteins)
  Endoplasmic reticulum (packages proteins)

b. **Ions associated with plasma membrane**—4 points maximum

(1 point each)

Ions	Description
Selectively permeable to K^+ ions	(Concentrated on the inside of the cell)
Impermeable to Na^+ ions	(Concentrated on the outside of the cell)
Build up of positively charged ions outside of the membrane	(Inside is negatively charged)
Sodium-potassium pump	(Maintains the ion gradient)

c. **Role of the membrane**—4 points maximum

(1 point each)
Nerve impulses are electrochemical (associated with ion and electrical changes in plasma membrane)
Action potential (change in the membrane potential)
Resting stage (voltage charge is –70 millivolts)
Depolarization (Na^+ moves into the cell down its cell gradient)
Repolarization (K^+ ions move out of the cell down their cell gradient)
Sodium channels (membrane channel that is voltage-gated)
Potassium channels (membrane channel that is voltage-gated)
Sodium-potassium pump (membrane protein that maintains ion gradient)

ESSAY 1

A nerve cell is fundamentally similar in structure and function to other somatic cells. Like other cells, a nerve cell consists of a bilipid layer made up of phospholipids and proteins. This membrane is semi-permeable; it regulates the passage of certain ions across its membrane. The cell body of a nerve cell also contains cellular structures such as a nucleus, which regulates the activities of the cell, and ribosomes, which make proteins.

The plasma membrane of a resting neuron is selectively permeable to K^+ ions and impermeable to Na^+ ions. Because of this selective permeability, the K^+ ions are concentrated on the inside of the cell and are able to slowly diffuse outward, while the Na^+ ions are concentrated on the outside of the cell. This leads to a build up of positively-charged ions outside of the neuronal membrane. The inside of the plasma membrane is therefore negatively-charged. These chemical gradients are maintained by a sodium-potassium pump. This pump actively moves three Na^+ ions out of the cell for every two K^+ ions brought into the cell.

The neuronal membrane plays an important role in the conduction of a nerve impulse. Nerve impulses are electrochemical. This means, the forces that cause ions to move across a membrane are both a concentration gradient and an electrical gradient. The membrane is associated with two chemicals (Na^+ ions and K^+ ions) as well as a change in the voltage charge. When a nerve cell is undisturbed the membrane is said to be in a resting stage. The membrane is polarized and the voltage charge is –70 millivolts. During a nerve impulse, there is a change in the membrane permeability and Na^+ ions rush into the cell and the inside becomes more positively-charged. The membrane is now said to be depolarized. Na^+ move into the cell down its concentration gradient. Next, the Na^+ channels close and K^+ channels open and K^+ ions move out of the cell. The cell is now said to be repolarized. The original ion concentration is reestablished by the sodium-potassium pump.

ESSAY CHECKLIST QUESTION 2

I. **Adaptational Problems**—4 points maximum

   (1 point each)

   Plants required water on land

   Plants required specialized plant organs (roots, stems, and leaves)

   Plants needed a method to carry out sexual reproduction on land

   Plants needed to develop physiological adaptations that would meet certain specific challenges (need for moisture via transpiration/need to withstand high temperatures via evaporation)

II. **Structural adaptations to acquire and retain water in land plants**—4 points maximum

   First three structures only

(1 point each)	**(1 point each)**
Structure	**Function described**
Cuticles	Retard water loss
Stomata	Pores in leaf that regulate water and gas exchange
Roots	Structures that absorb water and minerals
Root hairs	Extensions of the root that increase the surface area for water absorption
Xylem	Transport water and minerals throughout plant

III. Other structural adaptations—4 points maximum

First two structures only

(1 point each)	**(1 point each)**
Structure	**Function described**
Cuticles	Retards water loss
Spores	Prevent desiccation
Stomata	Pores in leaf that regulate water and gas exchange
Roots	Structures that absorb water and minerals
Root hairs	Extensions of the root that increase the surface area for water absorption
Xylem	Transport water and minerals throughout plant
Phloem	Transport nutrients
Spores	Resist desiccation; dormancy
Leaves	Contain photosynthetic cells
Stems	Contain conductive system and support plant
Seed/seedcoat	Protection of embryo, food
Flowers	Facilitate sexual reproduction

ESSAY 2

There were several adaptational problems that had to be solved before plants could survive on land without water. First, plants needed to develop different methods to retain water while on land. Second, plants required specialized organs such as roots and leaves. Third, plants needed to develop a method of conducting sexual reproduction in a terrestrial environment. Fourth, plants developed physiological adaptations to deal with their new environment. These include finding ways of keeping enough moisture and withstanding extreme temperatures.

In order to acquire and retain water, plants developed waxy cuticles on their outer surfaces, which prevent desiccation. They also developed stomates on their leaf surfaces, which controlled water loss (transpiration). Plants also contain root hairs, which absorb water and minerals. Plants also developed other structures to survive on land. Some cells specialized and became leaves with cells that carry out photosynthesis. Other cells specialized and became stems with vascular tissues, which include xylem (which conduct water) and phloem (which conduct food).

ESSAY CHECKLIST QUESTION 3

a. Type of Mutation—4 points maximum

(1 point each)
Base substitution (involves one nucleotide being replaced by another)
Change in DNA/codon of mRNA draws new tRNA molecule
New tRNA carries wrong amino acid, altering polypeptide
Change in polypeptide in turn alters hemoglobin protein
Distorted red blood cell cannot efficiently carry oxygen

b. **Gel Electrophoresis**—5 points maximum

How it works—3 points maximum
(1 point each)

Apparatus	DNA is put into wells of an agarose gel with a buffer
Electricity	Electrical potential (electrical charge moves fragments)
Charge	Negatively charged fragments move toward the positive pole
Size/Molecular weight	Smaller fragments move faster across the gel

How it can be applied to identify the mutant gene—2 points maximum
(1 point each)

Compare the normal hemoglobin to the abnormal hemoglobin

Use restriction enzymes to cut the two hemoglobin proteins into several fragments

Normal hemoglobin can be used as a marker

Fragments of the two proteins should be identical except for the fragment containing the abnormal gene

c. **Punnett Square**—2 points maximum

A normal female who does not carry the X-linked allele mates with a male homozygous for the disease:

HH → hh → All the offspring are Hh (heterozygotes)

They are all carriers (sickle cell trait)

ESSAY 3

Sickle cell anemia is a disease in which the red blood cells have an abnormal shape because of a base substitution. Base substitution involves an error in DNA replication in which one nucleotide is replaced by another nucleotide. This causes the codon of an mRNA to contain an incorrect base. The codon therefore matches up with the anticodon of a different tRNA. This tRNA carries a different amino acid. The change in the amino acid alters the polypeptide. The polypeptide, in turn, alters the hemoglobin protein. In this case, the distorted red blood cell cannot efficiently carry oxygen.

Biologists must have determined the nature of the hemoglobin mutation by comparing the normal hemoglobin to the abnormal hemoglobin using the technique gel electrophoresis. Gel electrophoresis identifies the difference between two molecules by examining the different rates each molecule moves across the gel. Substances move across a gel according to their molecular weight. For example, smaller fragments move faster than larger fragments. The biologists must have placed fragments of the two proteins on an agarose gel and used the normal hemoglobin as the marker—the source of comparison to the abnormal hemoglobin. (A restriction enzyme was used to cut the two proteins into several fragments prior to loading the gel). The fragments of the two proteins should have been identical except for the fragment that contained the abnormal gene.

If a normal non-carrier female mates with a male who is homozygous for the disease, these are the results using a Punnett square:

	H	H
h	Hh	Hh
h	Hh	Hh

All of the offspring would be heterozygotes. For sickle cell anemia, these offspring would be carriers.

ESSAY CHECKLIST QUESTION 4

a. **Krebs cycle, the electron transport chain, oxidative phosphorylation as aerobic processes**—2 points maximum

> They require oxygen
> They cannot occur under anaerobic conditions
> Glycolysis can occur under both anaerobic and aerobic conditions

b. **Site of each step**—3 points maximum

Stage	Site
Krebs cycle	Mitochondrial matrix
Electron transport chain	Along the inner mitochondrial membrane
Oxidative phosphorylation	As hydrogens move from the intermembrane space to the mitochondrial matrix

c. **Role of NADH and FADH$_2$ in aerobic respiration**—5 points maximum

> **(1 point each)**
> NADH and FADH$_2$ are electron/hydrogen ion carriers
> Coenzymes NAD$^+$ and FAD$^+$ accept electrons (and hydrogen) to form NADH and FADH$_2$
> NADH and FADH$_2$ shuttle electrons to the electron transport chain
> The hydrogens dissociate into electrons ions and electrons
> The electrons are passed down the electron transport chain to form water
> The hydrogen ions are pumped out into the intermembrane space and cross back to produce ATP

Essay 4

a. The Krebs cycle, electron transport chain, and oxidative phosphorylation are all part of aerobic respiration. During aerobic respiration, glucose is completely converted to CO_2, ATP, and water. These steps are considered aerobic processes because they cannot occur under anaerobic conditions; they require oxygen. Glycolysis, on the other hand, can occur under both aerobic and anaerobic conditions.

b. These three stages of aerobic respiration occur in different parts of the mitochondria. The Krebs cycle occurs in the mitochondrial matrix. Two acetyl CoA enter the Krebs cycle and produce NADH, $FADH_2$, ATP, and CO_2. The products of the Krebs cycle (NADH and $FADH_2$), are sent to the electron transport chain. The electron transport chain occurs along the inner mitochondrial membrane. The final stage of aerobic respiration—oxidative phosphorylation—occurs as hydrogens move across from the intermembrane space to the mitochondrial matrix.

c. NADH and $FADH_2$ are hydrogen ion and electron carriers. When the coenzymes NAD^+ and FAD^+ accept hydrogens they form NADH and $FADH_2$. NADH and $FADH_2$ shuttle the hydrogens to the electron transport chain, which have the potential to make a lot of ATP. Once the carriers drop off the hydrogens at the electron transport chain, the hydrogens dissociate into hydrogen ions and electrons. The electrons are sent down the electron transport chain and eventually make water. Meanwhile, the hydrogen ions are pumped out into the intermembrane space. When they cross back into the mitochondrial matrix they produce ATP.

19

The Princeton Review
AP Biology
Practice Test 2

BIOLOGY

Three hours are allotted for this examination: 1 hour and 20 minutes for Section I, which consists of multiple-choice questions, and 1 hour and 40 minutes for Section II, which consists of essay questions.

SECTION I

Time—1 hour and 20 minutes
Number of questions—100
Percent of total grade—60

Section I of this examination contains 100 multiple-choice questions, followed by 15 multiple-choice questions regarding your preparation for this exam. Please be careful to fill in only the ovals that are preceded by numbers 1 through 115 on your answer sheet.

General Instructions

INDICATE ALL YOUR ANSWERS TO QUESTIONS IN SECTION I ON THE SEPARATE ANSWER SHEET ENCLOSED. No credit will be given for anything written in this examination booklet, but you may use the booklet for notes or scratchwork. After you have decided which of the suggested answers is best, COMPLETELY fill in the corresponding oval on the answer sheet. Give only one answer to each question. If you change an answer, be sure that the previous mark is erased completely.

Example: Sample Answer

Chicago is a

(A) state
(B) city
(C) country
(D) continent
(E) village

Many candidates wonder whether or not to guess the answer to questions about which they are not certain. In this section of the examination, as a correction for haphazard guessing, one-fourth of the number of questions you answer incorrectly will be subtracted from the number of questions you answer correctly. It is improbable, therefore, that mere guessing will improve your score significantly; it may even lower your score, and it does take time. If, however, you are not sure of the correct answer but have some knowledge of the question and are able to eliminate one or more of the answer choices as wrong, your chance of getting the right answer is improved, and it may be to your advantage to answer such a question.

Use your time effectively, working as rapidly as you can without losing accuracy. Do not spend too much time on questions that are difficult. Go on to other questions and come back to the difficult ones later if you have time. It is not expected that everyone will be able to answer all the multiple-choice questions.

BIOLOGY

SECTION I

Time—1 hour and 20 minutes

<u>Directions:</u> Each of the questions or incomplete statements below is followed by five suggested answers or completions. Select the one that is best in each case and then fill in the corresponding oval on the answer sheet.

1. In general, animal cells differ from plant cells in that animal cells have

   (A) an endoplasmic reticulum and membrane-bound organelles
   (B) a cell wall made of cellulose
   (C) lysosomes
   (D) large vacuoles, that store water
   (E) centrioles, that are involved in cell division

2. In order for plants to conquer land they had to evolve all of the following adaptations EXCEPT

   (A) flagellated sperm cells that swim to fertilize an egg
   (B) a waxy cuticle layer on their outer surfaces
   (C) specialized structures that permit gas exchange
   (D) specialized organs that anchor the plant on land
   (E) vascular tissues that allow for the transport of needed nutrients and water

3. In a diploid organism with the genotype AaBbCCDDEE, how many genetically distinct kinds of gametes would be produced?

   (A) 4
   (B) 8
   (C) 16
   (D) 32
   (E) 64

4. A cell from the leaf of the aquatic plant *Elodea* was soaked in a 15 percent sugar solution and its contents soon separated from the cell wall and formed a mass in the center of the cell. All of the following statements are true about this event EXCEPT

   (A) The vacuole lost water and became smaller.
   (B) The space between the cell wall and the cell membrane expanded.
   (C) The large vacuole contained a solution with much lower osmotic pressure than that of the sugar solution.
   (D) The concentration of solutes in the extracellular environment is hypertonic with respect to the cell's interior.
   (E) The sugar solution passed freely through the cell wall but not the cell membrane.

5. Under favorable conditions, bacteria divide every 20 minutes. If a single bacterium replicated according to this condition, how many bacterial cells would one expect to find at the end of three hours?

   (A) 32
   (B) 64
   (C) 128
   (D) 256
   (E) 512

6. A chemical agent is found to denature all enzymes in the synaptic cleft. What effect will this agent have on acetylcholine?

   (A) Acetylcholine will not be released from the presynaptic membrane.
   (B) Acetylcholine will not bind to receptor proteins on the postsynaptic membrane.
   (C) Acetylcholine will not diffuse across the cleft to the postsynaptic membrane.
   (D) Acetylcholine will be inactivated by the chemical agent.
   (E) Acetylcholine will not be degraded in the synaptic cleft.

GO ON TO THE NEXT PAGE

7. The length of DNA molecules is measured in base pairs. The base composition of DNA varies from one species to another. Which of the following ratios would you expect to remain constant in the DNA?

(A) Cytosine / Adenine
(B) Pyrimidine / Purine
(C) Adenine / Guanine
(D) Guanine / Deoxyribose
(E) Thymine / Guanine

8. The use of radioactive iodine as a tracer element in the study of human metabolic rate has shown that iodine

(A) alters gene expression
(B) regulates calcium metabolism
(C) binds to a specific receptor on the cell membrane of a thyroid gland cell
(D) is a hormone that lowers glucose levels in the blood
(E) is a lipid-soluble hormone that diffuses across a membrane

9. In reptile eggs, the extraembryonic membrane that functions in excretion and respiration is the

(A) amnion
(B) chorion
(C) allantois
(D) yolk sac
(E) placenta

10. Consider the following enzyme pathway:

$$A \xrightarrow{1} B \xrightarrow{2} C \xrightarrow{3} D \xrightarrow{4} E \xrightarrow{5} F$$
$$C \xrightarrow{6} X \xrightarrow{7} Y$$

An increase in substance F leads to the inhibition of enzyme 3. All of the following are results of the process EXCEPT

(A) an increase in substance X.
(B) increased activity of enzyme 6.
(C) decreased activity of enzyme 4.
(D) increased activity of enzyme 5.
(E) a decrease in substance D.

11. The liver is a vital organ that performs all of the following functions EXCEPT

(A) stores amino acids that were absorbed in the capillaries of the small intestine
(B) detoxifies harmful substances such as alcohol or certain drugs
(C) synthesizes bile salts which emulsify lipids
(D) breaks down peptides into amino acids
(E) stores fatty acids that were absorbed by the lacteals of the lymphatic vessels

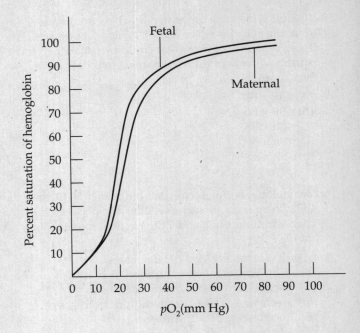

12. The graph above shows the oxygen dissociation curves of maternal hemoglobin and fetal hemoglobin. Based on the graph, it can be concluded that

(A) fetal hemoglobin surrenders O_2 more readily than maternal hemoglobin
(B) the dissociation curve of fetal hemoglobin is to the right of maternal hemoglobin
(C) fetal hemoglobin has a higher affinity for O_2 than does maternal hemoglobin
(D) fetal and maternal hemoglobin differ in structure
(E) fetal hemoglobin is converted to maternal hemoglobin

GO ON TO THE NEXT PAGE

13. In minks, the gene for brown fur (B) is dominant over the gene for silver fur (b). Which set of genotypes represents a cross that could produce offspring with silver fur from parents that both have brown fur?

(A) BB × BB
(B) BB × Bb
(C) Bb × Bb
(D) Bb × bb
(E) bb × bb

14. Hemoglobin is a molecule that binds to both O_2 and CO_2. There is an allosteric relationship between the concentrations of O_2 and CO_2. Hemoglobin's affinity for O_2

(A) decreases as blood pH decreases
(B) increases as H^+ concentration increases
(C) increases in exercising muscle tissue
(D) decreases as CO_2 concentration decreases
(E) increases as HCO_3^- increases

15. Nitrogen from the atmosphere must be incorporated into living organisms to make proteins. Which of the following plants is a vehicle for organisms that add nitrates into the soil?

(A) Rice
(B) Lima bean
(C) Rose
(D) Venus flytrap
(E) Corn

16. In snapdragon plants which display intermediate dominance, the allele C^R produces red flowers and C^W produces white flowers. If a homozygous red-flowered snapdragon is crossed with a homozygous white-flowered snapdragon, the percent ratio of the offspring will be

(A) 100% red
(B) 100% pink
(C) 50% red and 50% white
(D) 50% red and 50% pink
(E) 25% red, 50% pink, and 25% white

17. The principal components of a virus are

(A) DNA and proteins
(B) nucleic acid and protein coat
(C) DNA and cell membrane
(D) RNA and cell wall
(E) nucleic acid and cell membrane

18. One characteristic that flagellates, ciliates, sporozoans, and green algae have in common is that they

(A) undergo alteration of generation
(B) form spores
(C) use flagella for motility
(D) are autotrophic
(E) are microscopic, single-celled, eukaryotes

19. The scientific name for the fruit fly is *Drosophila melanogaster*. The word *Drosophila* refers to the classification group known as

(A) species
(B) genus
(C) class
(D) family
(E) phylum

20. In humans, fertilization normally occurs in the

(A) ovary
(B) fallopian tube
(C) uterus
(D) placenta
(E) vagina

21. The development of an egg without fertilization is known as

(A) meiosis
(B) cloning
(C) embryogenesis
(D) vegetative propagation
(E) regeneration

GO ON TO THE NEXT PAGE

22. All of the following are examples of hydrolysis EXCEPT

(A) conversion of fats to fatty acids and glycerol
(B) conversion of proteins to amino acids
(C) conversion of starch to simple sugars
(D) conversion of pyruvic acid to acetyl CoA
(E) conversion of proteins to dipeptides

23. In cells, which of the following can catalyze reactions involving hydrogen peroxide, provide cellular energy, and make proteins, in that order?

(A) Peroxisomes, mitochondria, and ribosomes
(B) Peroxisomes, mitochondria, and lysosomes
(C) Peroxisomes, mitochondria, and Golgi apparatus
(D) Lysosomes, chloroplasts, and ribosomes
(E) Smooth endoplasmic reticulum, mitochondria, and ribosomes

24. All of the following play an important role in regulating respiration in humans EXCEPT

(A) an increase in the amount of CO_2 in the blood.
(B) a decrease in the amount of O_2 concentration in the blood.
(C) a decrease in the plasma pH level.
(D) strenuous exercise.
(E) an increase in H^+ levels.

25. The primary site of glucose reabsorption is the

(A) glomerulus
(B) proximal convoluted tubule
(C) loop of Henle
(D) collecting duct
(E) distal convoluted tubule

Questions 26 and 27 refer to the graph.

The graph below shows the growth curve of a bacterial culture.

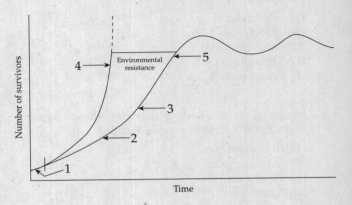

26. Which of the following represents the carrying capacity of the environment?

(A) 1
(B) 2
(C) 3
(D) 4
(E) 5

27. Which of the following shows the exponential growth phase of the population?

(A) 1
(B) 2
(C) 3
(D) 4
(E) 5

28. Which of the following hormones is incorrectly paired with its respective function?

(A) Thyroxine…increases the rate of metabolism
(B) Insulin…decreases storage of glycogen
(C) Vasopressin…stimulates water reabsorption in the kidney
(D) Epinephrine…increases blood sugar levels and heart rate
(E) Growth hormone…stimulates muscle growth

GO ON TO THE NEXT PAGE

29. Metafemale syndrome, a disorder in which a female has an extra X chromosome, is the result of nondisjunction. The failure in oogenesis that could produce this would occur in

    (A) Prophase I
    (B) Metaphase I
    (C) Metaphase II
    (D) Telophase I
    (E) Anaphase II

30. In plants, the tendency of climbing vines to twine their tendrils around a trellis is called

    (A) thigmotropism
    (B) hydrotropism
    (C) phototropism
    (D) geotropism
    (E) chemotropism

31. Females with Turner's syndrome have a high incidence of hemophilia, a recessive, X-linked trait. Based on this information, it can be inferred that females with this condition

    (A) have an extra X
    (B) have an extra Y
    (C) lack an X
    (D) have red blood cells that clump
    (E) have an abundance of platelets

32. When a retrovirus inserted its DNA into the middle of a bacterial gene, it altered the normal reading frame by one base pair. This type of mutation is called

    (A) duplication
    (B) translocation
    (C) inversion
    (D) frameshift mutation
    (E) lethal mutation

33. High levels of estrogen from maturing follicles inhibit the release of gonadotropin releasing hormone (GnRH). Which of the following endocrine structures produces GnRH?

    (A) Anterior pituitary
    (B) Posterior pituitary
    (C) Hypothalamus
    (D) Pineal gland
    (E) Ovary

34. At times, freshwater communities undergo changes in which the availability of nutrients increases to the extent that an overabundance of green algae suffocates the community. This trend toward excess nutrients in freshwater bodies is called

    (A) succession
    (B) eutrophication
    (C) evolution
    (D) greenhouse effect
    (E) lake turnover

35. The principle inorganic compound found in living things is

    (A) carbon
    (B) oxygen
    (C) water
    (D) glucose
    (E) carbon dioxide

36. Kangaroo rats are better able to concentrate urine than humans are. It would be expected that compared to the nephrons of human kidneys, the nephrons of kangaroo-rat kidneys would have

    (A) thicker walls, which are impermeable to water
    (B) shorter loops of Henle
    (C) longer loops of Henle
    (D) shorter collecting ducts
    (E) longer proximal convoluted tubules

GO ON TO THE NEXT PAGE

37. All of the following are modes of asexual reproduction EXCEPT

(A) sporulation
(B) fission
(C) budding
(D) cloning
(E) meiosis

38. The moist skin of earthworms, the lenticels of plants, and the spiracles of grasshoppers are all associated with the process of

(A) excretion
(B) respiration
(C) circulation
(D) digestion
(E) reproduction

39. Locomotion in annelids is accomplished through the interaction of muscles and

(A) an exoskeleton
(B) paired setae
(C) tracheids
(D) jointed appendages
(E) pseudopods

40. All of the following are examples of connective tissue EXCEPT

(A) ligaments
(B) muscle
(C) blood
(D) cartilage
(E) bone

41. In most plants, germination is triggered by the presence of

(A) water, oxygen, and soil
(B) light, water, and soil
(C) carbon dioxide, water, and soil
(D) oxygen, water, and temperature
(E) carbon dioxide and light

42. Which of the following statements best describes a pyramid of energy?

(A) A net gain occurs as energy is transferred from one organism to another.
(B) The total energy in plants is less than that in herbivores.
(C) The first trophic level is at the top of a pyramid.
(D) The total mass of carnivores is more than the total mass of plants.
(E) Each smaller trophic level possesses less available energy than the previous level.

43. If a forest of fir, birch, and white spruce trees was devastated by fire, which of the following would most likely happen?

(A) Only animal life would continue to inhabit the region.
(B) Secondary succession would begin to occur.
(C) Only tough grasses would appear.
(D) The number of species would stabilize as the ecosystem matures.
(E) This forest would no longer support living things.

44. Which of the following processes from the list below occur in the cytoplasm of a eukaryotic cell?

I. DNA replication

II. Transcription

III. Translation

(A) I only
(B) III only
(C) I and III only
(D) II and III only
(E) I, II, and III

45. Crossing-over during meiosis permits scientists to determine

(A) the chance for variation in zygotes
(B) the rate of mutations
(C) chromosome mapping
(D) which traits are dominant or recessive
(E) which traits are masked

GO ON TO THE NEXT PAGE

46. Which of the following statements describes an activity that is part of the nitrogen cycle?

   (A) Legume plants release water into the atmosphere through the process of transpiration.
   (B) Green plants assimilate nitrogen in the form of ammonia.
   (C) Soil bacteria convert ammonia into minerals usable by autotrophs.
   (D) Bacteria return nitrogen that is locked up in urea to the nitrogen cycle.
   (E) Legume plants only assimilate nitrogen through the activity of small microorganisms.

47. The results of the process of cloning are most similar to the results of the process of

   (A) gametogenesis
   (B) fertilization
   (C) pollination
   (D) meiosis
   (E) mitosis

48. Three distinct bird species, flicker, woodpecker, and elf owl, all inhabit a large cactus, *Cereus giganteus*, in the desert of Arizona. Since competition among these birds rarely occurs, the most likely explanation for this phenomenon is that these birds

   (A) have a short supply of resources
   (B) have different ecological niches
   (C) do not live together long
   (D) are unable to breed
   (E) do not share the same habitat

49. Lampreys attach to the skin of lake trout and absorb nutrients from its body. This relationship is an example of

   (A) commensalism
   (B) parasitism
   (C) mutualism
   (D) gravitropism
   (E) thigmotropism

50. The nucleotide sequence of a DNA molecule is 5'-C-A-T-3'. A mRNA molecule with a complementary codon is transcribed from the DNA in the process of protein synthesis a tRNA pairs with a mRNA codon. What is the nucleotide sequence of the tRNA anticodon?

   (A) 5'-G-T-A-3'
   (B) 5'-G-U-A-3'
   (C) 5'-C-A-U-3'
   (D) 5-'U-A-C-3'
   (E) 5'-G-U-G-3'

51. Viruses are considered an exception to the cell theory because they

   (A) are not independent organisms
   (B) have only a few genes
   (C) move about via their tails
   (D) have evolved from ancestral protists
   (E) are multinucleated

52. All of the following organs in the digestive system secrete digestive enzymes EXCEPT the

   (A) mouth
   (B) stomach
   (C) gall bladder
   (D) small intestine
   (E) pancreas

53. Memory loss would most likely be due to a malfunction of which part of the brain?

   (A) Medulla
   (B) Cerebellum
   (C) Cerebrum
   (D) Pons
   (E) Hypothalamus

GO ON TO THE NEXT PAGE

54. In nonplacental mammals, the embryo obtains its food from the

(A) ovary
(B) uterus
(C) oviduct
(D) yolk sac
(E) allantois

55. The sequence of amino acids in hemoglobin molecules of humans is more similar to the hemoglobin of chimpanzees than it is to the hemoglobin of dogs. This similarity suggests that

(A) humans and dogs are more closely related than humans and chimpanzees
(B) humans and chimpanzees are more closely related than humans and dogs
(C) humans are related to chimpanzees but not to dogs
(D) humans and chimpanzees are closely analogous
(E) the hemoglobin molecule of all three organisms did not have intervening sequences

56. According to the heterotroph hypothesis, which event had to occur before oxygen filled the atmosphere?

(A) Heterotrophs had to remove carbon dioxide from the air
(B) Autotrophs, which make their own food, had to evolve
(C) Heterotrophs had to evolve
(D) Autotrophs had to convert atmospheric nitrogen to nitrate
(E) Heterotrophs had to carry out their metabolic activity in the presence of oxygen

57. Which structure in an earthworm has a function similar to that of the alveoli of a human?

(A) Malphigian tubules
(B) Nephridia
(C) Chitinous exoskeleton
(D) Gills
(E) Skin

GO ON TO THE NEXT PAGE

Directions: Each group of questions consists of five lettered headings followed by a list of numbered phrases or sentences. For each numbered phrase or sentence, select the one heading that is most closely related to it and fill in the corresponding oval on the answer sheet. Each heading may be used once, more than once, or not at all in each group.

Questions 58–61

(A) Cortisol
(B) Oxytocin
(C) Progesterone
(D) Hypothalamus
(E) Parathyroid hormone

58. Regulates bone growth

59. Elevates blood sugar

60. Maintains uterine endometrium

61. Secretes hormones that travel to the anterior pituitary

Questions 62–65

(A) Insight learning
(B) Operant conditioning
(C) Imprinting
(D) Classical conditioning
(E) Circadian rhythm

62. Bees that regularly feed on blue paper settle on blue paper even though it lacks food

63. Hatched goslings become attached to the first moving object they see

64. A child learns not to touch a hot stove

65. Opening and closing of stomates, independent of light and darkness

Questions 66–68

(A) Mesoderm
(B) Ectoderm
(C) Endoderm
(D) Coelom
(E) Amnion

66. Germ layer that gives rise to the skin

67. Germ layer that gives rise to the epithelial lining of the gastrointestinal tract and its outgrowths

68. Germ layer that gives rise to cartilage, bone, and other connective tissues

GO ON TO THE NEXT PAGE

(A) Taiga
(B) Tundra
(C) Temperate forest
(D) Grassland
(E) Tropical rain forest

69. Contains soil that has the highest rate of leaching of nutrients

70. Characterized by low temperatures and a short growing season

71. Animal life includes black bears, moose, and wolves

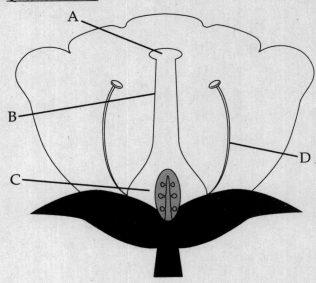

72. Embryo sporophyte develops within this structure

73. Structure that has a similar function to that of the human testes

74. Sticky, pollen-trapping structure

75. Meiosis occurs within this structure to produce microspores

GO ON TO THE NEXT PAGE

(A)

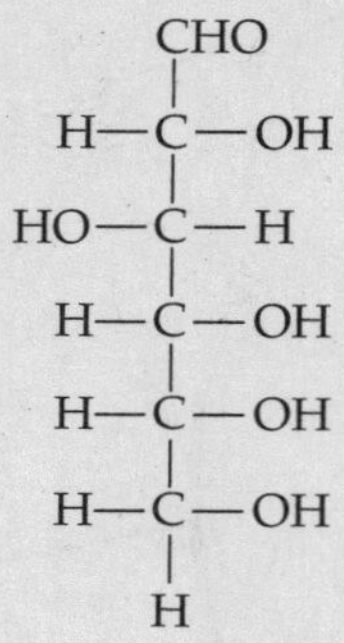

```
        CHO
    H — C — OH
   HO — C — H
    H — C — OH
    H — C — OH
    H — C — OH
        |
        H
```

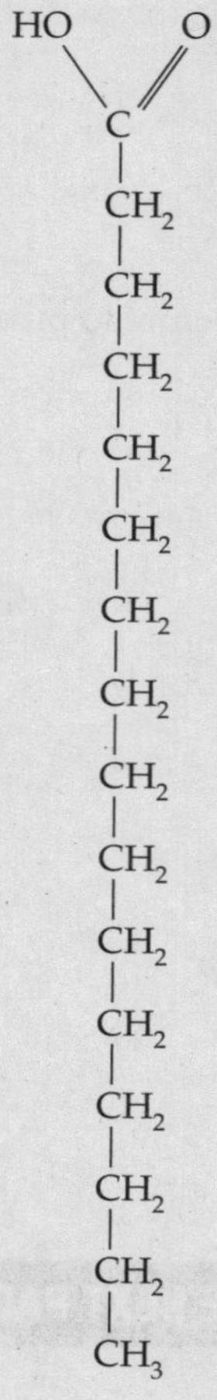

(D)

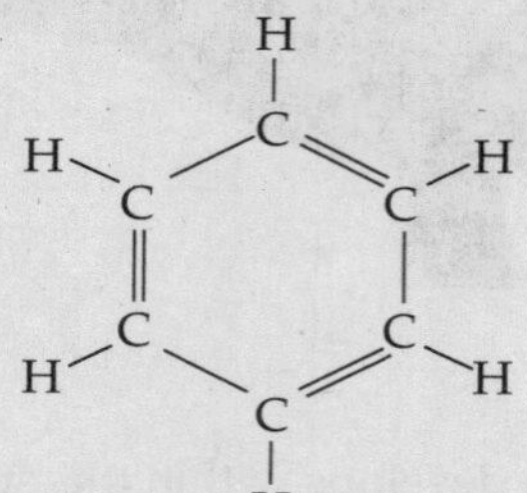

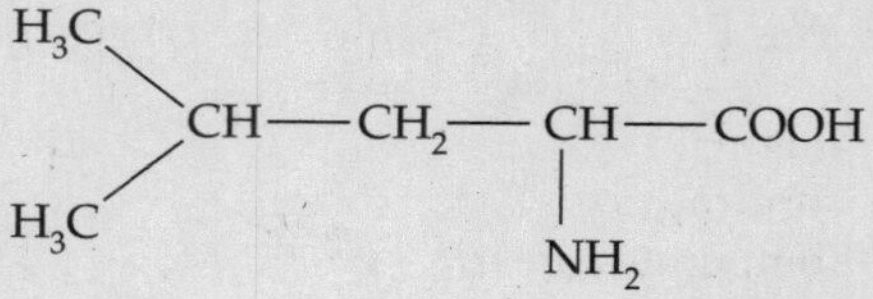

```
H₃C
    \
     CH — CH₂ — CH — COOH
    /                |
H₃C               NH₂
```

(B)

```
        H
        |
        C
      /   \\
  H—C       C—H
   ‖         |
  H—C       C—H
      \\   /
        C
        |
        H
```

(E)

```
                          NH₂
                          |
                          C
                        /   \\
                    N         N
                    ‖          \\
             O⁻    HC          C   C
             |       \\        /  \\ ‖
    ⁻O — P — O        N       C    C
             ‖                  \\  /
             O                   N
                                 |
              O — CH₂
             /       \
            /         \
           O
          H         H
        H             H
          |           |
         OH          OH
```

(C)

```
  HO      O
    \    ‖
     C
     |
     CH₂
     |
     CH₂
     |
     CH₂
     |
     CH₂
     |
     CH₂
     |
     CH₂
     |
     CH₂
     |
     CH₂
     |
     CH₂
     |
     CH₂
     |
     CH₂
     |
     CH₂
     |
     CH₂
     |
     CH₃
```

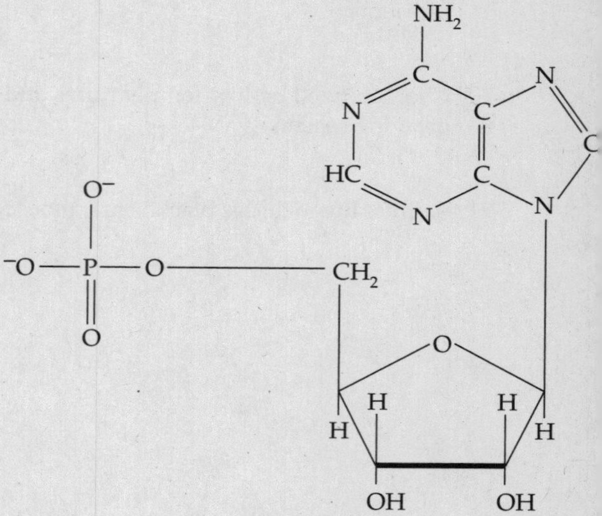

76. Subunits of nucleic acids

77. A building block of phospholipids

78. An important energy source central to many metabolic processes

79. A building block of proteins

GO ON TO THE NEXT PAGE

Questions 80–82 refer to the following experiment.

A group of 100 female *Daphnia*, small crustaceans known as water fleas, were placed in one of three culture jars of different sizes to determine their reproductive rate. The graph below shows the average number of offspring produced per female each day in each jar of pond water.

Key: (A) Water fleas in a 1-liter jar of pond water
(B) Water fleas in a 1/2-liter jar of pond water
(C) Water fleas in a 1/4-liter jar of pond water

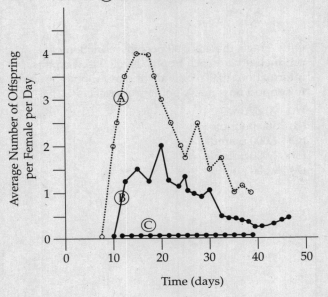

Time (days)

80. What is the total number of offspring produced in the 0.5-liter jar on the twentieth day, assuming all survive?

(A) 2
(B) 4
(C) 50
(D) 200
(E) 400

81. Based on the graph, what is the limiting factor in the reproductive rate of the female water fleas?

(A) Vapor pressure
(B) Food
(C) Temperature
(D) Density
(E) Competition

82. Which of the following statements is true concerning the results of the experiment?

(A) The water fleas in the 1-liter jar have a lower reproductive rate than the water fleas in the 0.5-liter jar.
(B) The reproductive rate of the water fleas in jars A and B are similar.
(C) The reproductive rate in the 0.25-liter jar changes because of a change in habitat.
(D) The water fleas in the 0.25-liter jar are infertile.
(E) The reproductive rate for the water fleas steadily decreases after 20 days.

GO ON TO THE NEXT PAGE

Questions 83–85 refer to the bar graph, which shows the relative biomass of four different populations of a particular food pyramid.

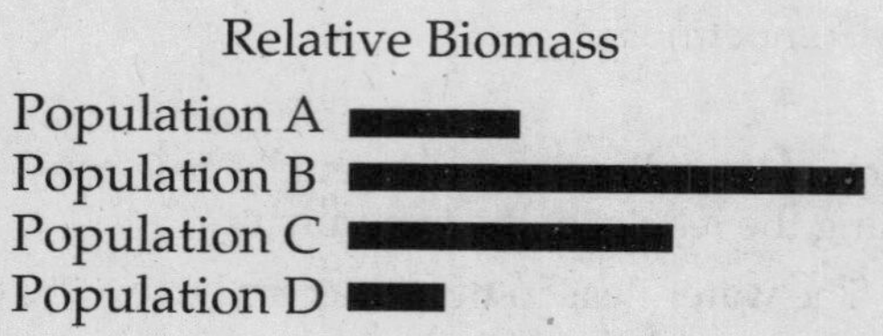

Relative Biomass

Population A ▬▬▬
Population B ▬▬▬▬▬▬▬▬▬▬
Population C ▬▬▬▬▬▬
Population D ▬▬

83. The largest amount of energy is available to

(A) population A
(B) population B
(C) population C
(D) population D
(E) It cannot be determined

84. Which of the following would be the most likely result if there was an increase in the number of organisms in population C?

(A) The biomass of population D will remain the same.
(B) The biomass of population B will decrease.
(C) The biomass of population C will steadily increase.
(D) The food source available to population C would increase.
(E) There would be an intense competition among the members of population C for the food source.

85. On average, there is a 90 percent reduction of productivity for each trophic level. Based on this information, 10,000 pounds of grass should be able to support how many pounds of crickets?

(A) 90 pounds
(B) 500 pounds
(C) 1,000 pounds
(D) 2,000 pounds
(E) 9,000 pounds

GO ON TO THE NEXT PAGE ➤

Questions 86–89 refer to the following illustration and information.

The cell cycle is a series of events in the life of a dividing eukaryotic cell. It consists of four stages: G_1, S, G_2, and M. The duration of the cell cycle varies from one species to another, and from one cell type to another. The G_1 phase varies the most. For example, embryonic cells can pass through the G_1 phase so quickly that it hardly exists, whereas neurons are arrested in the cell cycle and do not divide.

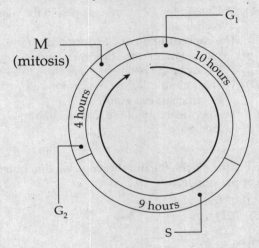

86. During which phase do chromosomes replicate?

(A) G_1
(B) S
(C) G_2
(D) M
(E) Cytokinesis

87. In mammalian cells, the first sign of prophase is the

(A) appearance of chromosomes
(B) separation of chromatids
(C) disappearance of the nuclear membrane
(D) replication of chromosomes
(E) crossing over of homologous chromosomes

88. Mitosis occurs in all of the following types of cells EXCEPT

(A) epidermal cells
(B) hair cells
(C) white blood cells
(D) pancreatic cells
(E) kidney cells

89. Since neurons are destined never to divide again, what can you conclude from this passage?

(A) These cells will go through cell division.
(B) These cells will be permanently arrested in the G_1 phase.
(C) These cells will be permanently arrested in the G_2 phase.
(D) These cells will quickly enter the S-phase.
(E) The duration of the cell cycle will be long.

GO ON TO THE NEXT PAGE →

Questions 90–93 refer to the graphs, which show the permeability of ions during an action potential in a ventricular contractile cardiac fiber. The action potential of cardiac muscle fibers resemble that of skeletal muscles.

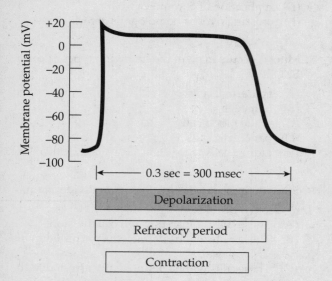

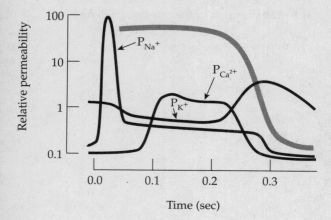

90. Based on the graph, the resting membrane potential of the muscle fibers is close to
    (A) −90 mV
    (B) −70 mV
    (C) 0 mV
    (D) +70 mV
    (E) +90 mV

91. Which of the following statements is true concerning the initial phase of depolarization?
    (A) Voltage-gated K^+ channels open in the plasma membrane.
    (B) The concentration of Ca^{2+} ions within the plasma membrane becomes more negative.
    (C) The membrane potential stays close to −40 mV.
    (D) There is a rapid inflow of Ca^{2+} ions along the electrochemical gradient.
    (E) The permeability of the sarcolemma to Na^+ ions increases.

92. In cardiac fibers, the duration of an action potential is approximately
    (A) 0.10 secs
    (B) 0.20 secs
    (C) 0.25 secs
    (D) 0.30 secs
    (E) 0.35 secs

93. One major difference between the action potential of cardiac muscle fibers and the action potential of skeletal muscle fibers is that in cardiac muscle fibers
    (A) the membrane is permeable to Na^+, not K^+
    (B) voltage-gated K+ channels open during depolarization, not repolarization
    (C) depolarization is prolonged compared to that in skeletal muscle fibers
    (D) the refractory period is shorter than that of skeletal muscle fibers
    (E) the speed of the contraction is faster than that of skeletal muscle fibers

GO ON TO THE NEXT PAGE

Questions 94–97 refer to the data below concerning the general animal body plan of five organisms.

	Acoelomate	Psudocoelomate	Coelomate	Protostomes	Deuterostomes	Radical Symmetry	Bilateral Symmetry
Organism 1						+	+
Organism 2	+						
Organism 3		+					+
Organism 4			+	+		+	
Orgainsim 5			+		+		+

Note: + indicates a feature presence in an organism.

94. The body plan associated with nematodes is Organism

(A) 1
(B) 2
(C) 3
(D) 4
(E) 5

95. The body plan associated with flatworms is Organism

(A) 1
(B) 2
(C) 3
(D) 4
(E) 5

96. Into what phylum should Organism 5 be placed?

(A) Platyhelminthes
(B) Nematoda
(C) Molluska
(D) Annelida
(E) Chordata

97. All of the following organisms exhibit bilateral symmetry EXCEPT

(A) hydra
(B) mollusks
(C) arthropods
(D) earthworms
(E) fish

GO ON TO THE NEXT PAGE

Questions 98–100 refer to the synthetic pathway of a pyrimidine, cytidine 5' triphosphate, CTP. This pathway begins with the condensation of two small molecules by the enzyme, aspartate transcarbamylase (ATCase).

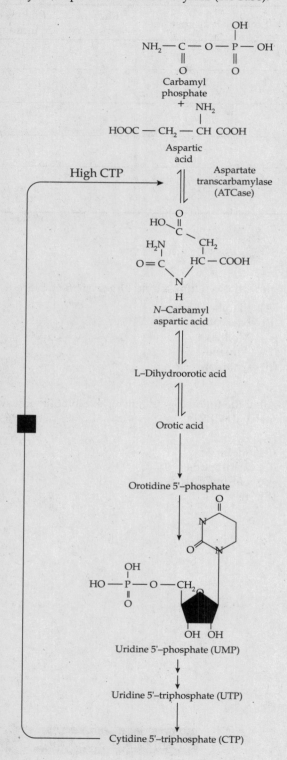

Regulation of CTP biosynthesis

98. When the level of CTP is low in a cell,
   (A) CTP is converted to ATCase
   (B) the metabolic traffic down the pathway increases
   (C) ATCase is inhibited, which slows down CTP synthesis
   (D) the final product of the pathway is reduced
   (E) CTP blocks the production of N-carbamyl aspartic acid

99. This enzymatic phenomenon is an example of
   (A) transcription
   (B) feedback inhibition
   (C) dehydration synthesis
   (D) photosynthesis
   (E) hydrolysis

100. The biosynthesis of cytidine 5'-triphosphate requires
   (A) a ribose sugar, a phosphate group, and a nitrogen base
   (B) a deoxyribose sugar, a phosphate group, and a nitrogen base
   (C) a ribose sugar, phosphate groups, and a nitrogen base
   (D) a deoxyribose sugar, phosphate groups, and a nitrogen base
   (E) a ribose sugar, phosphates, ATP, and a nitrogen base

STOP
IF YOU FINISH BEFORE TIME IS CALLED, YOU MAY CHECK YOUR WORK ON THIS SECTION.

DO NOT GO ON
UNTIL YOU ARE TOLD TO DO SO.

NO TEST MATERIAL ON THIS PAGE

BIOLOGY

SECTION II

Planning time—10 minutes

Writing time—1 hour and 30 minutes

You will have 10 minutes to read the exam questions. Spend this time reading through all of the questions, noting possible problem-solving approaches and otherwise planning your answers. It's fine to make notes on the green question insert, but be sure to write your answers and anything else that might be worth partial credit in the pink answer booklet—the graders will not see the green insert. After 10 minutes you will be told to break the seal on the pink Free-Response booklet and begin writing your answers in that booklet.

Answer all questions. Number your answer as the question is numbered below.

Answers must be in essay form. Outline form is NOT acceptable. Labeled diagrams may be used to supplement discussion, but in no case will a diagram alone suffice. It is important that you read each question completely before you begin to write.

1. Chlorophyll is one of a class of pigments that absorb light energy in photosynthesis.
   a. Relate the structure of chlorophyll to its function.
   b. Design an experiment to investigate the influence of sunlight on the activity of chlorophyll.
   c. Describe what information concerning the structure of chlorophyll could be inferred from your experiment.

2. Select **two** of the following three pairs of hormones and discuss the concept of negative feedback.
   a. Thyroid-stimulating hormone (TSH) and thyroxine
   b. Parathyroid hormone and calcitonin
   c. ACTH and cortisol

3. Describe the chemical nature of genes. Discuss the replication process of DNA in eukaryotic organisms. Name two types of gene mutations that could occur during replication.

4. Over the course of early evolution, organisms had to develop various methods to regulate internal fluids and excrete wastes. Discuss the problems faced by three organisms and how these problems were solved. In your discussion include structural adaptations and their functional significance.

END OF EXAMINATION

20

Answers and Explanations to Practice Test 2

1. **(E) is the correct answer.** Animal cells have centrioles that are responsible for cell division. (A) and (C): Both animal and plant cells have an endoplasmic reticulum, membrane-bound organelles and lysosomes. (B) and (D): Only plants have a cell wall made of cellulose and large vacuoles.

2. **(A) is the correct answer.** Plants were able to conquer land because they had a waxy cuticle to prevent water loss, stomates which allow for gas exchange, roots to anchor them, and vascular tissues to transport food and water. Land plants had to develop spores and gametes with a protective covering in order to reproduce.

3. **(A) is the correct answer.** There are four genetically distinct kinds of gametes that could be produced: ABCDE, AbCDE, aBCDE, and abCDE. Notice that the only alleles that vary are A and B ($2^2 = 4$).

4. **(D) is the correct answer.** If the contents of the cell separated from the cell wall then water was moving *out* of the cell. (B): This would cause the space between the cell wall and the cell membrane to expand. This also means that the concentration of solutes in the extracellular environment would therefore be *hypotonic* with respect to the cell's interior. (C): Since the vacuole is within the cell, it contains a solution with a much lower osmotic pressure than that of the sugar solution. (E): If water was moving out of the cell, the sugar solution would pass freely through the cell wall but not the cell membrane. (A): Lastly, since the fluid in the cell was hypertonic to the sugar solution, fluid was moving out of the vacuole and caused it to become smaller.

5. **(E) is the correct answer.** If bacteria divide every 20 minutes, you would produce 512 bacterial cells. One method would be to use the equation 2^x, where x equals the number of 20 minute intervals in three hours; $2^9 = 512$.

6. **(E) is the correct answer.** This question is testing your ability to associate what happens when enzymes are denatured and what would happen in the synaptic cleft. Acetylcholinesterase is an enzyme that degrades acetylcholine in the synaptic cleft. (A), (B) and (C): If acetylcholinesterase is denatured, acetylcholine will still be released from the presynaptic membrane and continue to diffuse across the synaptic cleft and bind to the postsynaptic membrane because acetylcholine is not degraded. (D): The chemical agent will have no affect on acetylcholine.

7. **(B) is the correct answer.** The ratio of purines to pyrimidines should be constant since purines always bind with pyrimidines, no matter which ones they may be.

8. **(C) is the correct answer.** Iodine is the major component of thyroid hormones T_3 and T_4 and can bind to a specific receptor on the cell membrane of a thyroid cell. (A): Iodine does not alter gene expression. (B): Calcitonin and parathyroid hormones regulate plasma calcium levels. (D): Insulin lowers the concentration of glucose in the blood. (E): Iodine is not a hormone.

9. **(C) is the correct answer.** The extraembryonic membrane that functions in respiration and excretion is the allantois. (A): The amnion is an extraembryonic membrane that protects the embryo. (B): The chorion is the outermost layer that surrounds the embryo. (D): The yolk sac provides nourishment for the embryo. (E): The placenta is the structure through which materials are exchanged between the fetus and mother.

10. **(D) is the correct answer.** If substance F leads to the inhibition of enzyme 3, then substances D and E and enzymes 3, 4, and 5 will be affected. The activity of enzyme 5 will be decreased, not increased.

11. **(D) is the correct answer.** The liver does not break down peptides into amino acids. It performs all of the following functions: (A): stores amino acids absorbed in the capillaries, (B): detoxifies harmful substances, (C): makes bile, and (E): stores fatty acids absorbed by the lacteals.

12. **(C) is the correct answer.** Based on the graph, fetal hemoglobin has a higher affinity for oxygen than maternal fetal hemoglobin. (A): Fetal hemoglobin does not give up oxygen more readily than maternal hemoglobin. (B): The dissociation curve of fetal hemoglobin is to the left of the maternal hemoglobin. (D): Fetal hemoglobin and maternal hemoglobin are different structurally, but you can't tell this from the graph. (E): Fetal hemoglobin does not convert to maternal hemoglobin.

13. **(C) is the correct answer.** The set of genotypes that represents a cross that could produce offspring with silver fur from parents that both have brown fur is Bb and Bb. Complete a Punnett square for this question. In order for the offspring to have silver fur, both parents must have the silver allele.

14. **(A) is the correct answer.** (B), (D) and (E): Hemoglobin's affinity for O_2 decreases as the concentration of H^+ increases (or the pH decreases) and as the concentration of CO_2 increases (or the concentration of HCO_3^- increases). (C): Hemoglobin's affinity for oxygen in tissue muscles does not increase during exercise.

15. **(B) is the correct answer.** Legumes (bean plants) are vehicles for organisms that add nitrates into the soil. An example of a legume is a lima bean.

16. **(B) is the correct answer.** Since snapdragon plants display intermediate dominance, the heterozygous phenotype is affected by the alleles of both homozygotes. If a homozygous red plant is crossed with a homozygous white plant, all offspring would be heterozygous and pink.

17. **(B) is the correct answer.** Viruses are made up of nucleic acid surrounded by a protein coat. They do not contain a cell wall, proteins, or cell membrane.

18. **(E) is the correct answer.** All of these organisms are protists—single-celled eukaryotic organisms. (A): These organisms do not undergo alternation of generation. (B): Sporozoans have spores. (C): Flagellates use flagella for motility. (D): Green algae are autotrophic.

19. **(B) is the correct answer.** The word *Drosophila* refers to the classification group known as genus. All organisms are given a scientific name consisting of a species name and a genus name. *Melanogaster* refers to the species name. The order of classification is: kingdom, phylum, class, order, family, genus, and species.

20. **(B) is the correct answer.** In humans, fertilization normally occurs in the fallopian tube. (A): The ovary is the female gonad that contains the eggs. (C): The uterus is the organ that houses the developing embryo. (D): The placenta is the structure that nourishes the embryo. (E): The vagina is the female organ in which sperm cells are deposited.

21. **(B) is the correct answer.** The development of an egg without fertilization is known as parthenogenesis. Parthenogenesis is a form of asexual reproduction found in insects and lizards. (A): Meiosis is a form of sexual reproduction that produces gametes. (C): Embryogenesis refers to the early stages of embryo development. (D): Vegetative propagation is a form of asexual reproduction by which plants produce identical offsprings from stem, leaves, or roots. (E): Regeneration is the restoration or new growth of a body part by an organism.

22. **(D) is the correct answer.** All of the following are examples of hydrolysis except the conversion of pyruvic acid to glucose. Hydrolysis is the breaking of a covalent bond by adding water. In all of the correct examples, complex compounds are broken down to simpler compounds. The conversion of pyruvic acid to acetyl CoA is an example of decarboxylation—a carboxyl group is removed as carbon dioxide and the 2-carbon fragment is oxidized.

23. **(A) is the correct answer.** Peroxisomes catalyze reactions that produce hydrogen peroxide, ribosomes are involved in protein synthesis, and mitochondria contain enzymes involved in cellular respiration. (B) and (D): Lysosomes are the sites of degredation; they contain hydrolytic enzymes, but do not produce hydrogen peroxide. (C): The Golgi apparatus sorts and packages substances that are destined to be secreted out of the cell. (E): The smooth endoplasmic reticulum is the site of lipid synthesis. Interestingly, O_2 concentration generally doesn't play an important role in regulating respiration.

24. **(B) is the correct answer.** Use process of elimination. The respiratory rate in humans will be affected by an increase in the amount of CO_2, a drop in pH levels, which is the same as an increase in hydrogen ion levels. Strenuous exercising will also modify the respiratory rate. O_2 concentration generally does not play an important role in regulating respiration.

25. **(B) is the correct answer.** The primary site of glucose reabsorption is the proximal convoluted tubule. (A): The glomerulus is a tuft of capillaries that filters fluid into the Bowman's capsule. (C): The loop of Henle is the site of salt reabsorption. (D): The collecting duct is the site in which urine is concentrated. (E): The distal convoluted tubule conducts filtrate and is adjacent to the collecting tubule of the nephron.

26. **(E) is the correct answer.** The carrying capacity is the maximum number of organisms of a given species that can be maintained in a given environment. Once a population reaches its carrying capacity, the number of organisms will fluctuate around it.

27. **(D) is the correct answer.** During the exponential growth phase of a population, population size doubles during each time interval. This part of the graph looks like a parabola.

28. **(B) is the correct answer.** Insulin decreases the level of blood glucose by increasing the storage of glycogen in muscles. The other hormones are correctly paired with their respective function.

29. **(E) is the correct answer.** The failure in oogenesis that could produce this syndrome would occur in anaphase I or anaphase II. Anaphase refers to the stage of meiosis in which chromatids separate from each other. If the chromosomes or chromatids fail to separate during anaphase, one egg cell will contain two X chromosomes, instead of one.

30. **(A) is the correct answer.** The tendency of climbing vines to twine their tendrils around a trellis is called thigmotropism. Thigmotropism refers to growth stimulated by contact with an object. (B): Hydrotropism refers to growth of a plant toward water. (C): Phototropism refers to growth toward light. (D): Geotropism refers to growth toward or against gravity. (E): Chemotropism refers to growth toward or away from a chemical substance.

31. **(C) is the correct answer.** Females with Turner's syndrome lack an X chromosome. If females with this syndrome have a high rate of hemophilia, they must not have the second X to mask the expression of the disease.

32. **(D) is the correct answer.** This type of mutation is called frameshift mutation. The insertion of DNA leads to a change in the normal reading frame by one base pair. The other answer choices refer to chromosomal aberrations. (A) Duplication is when an extra copy of a chromosome segment is introduced. (B): Translocation is when a segment of a chromosome moves to another chromosome. (C): Inversion is when a segment of a chromosome is inserted in the reverse orientation. (E): It is not known if this would be lethal or not.

33. **(C) is the correct answer.** The hypothalamus releases GnRH, which stimulates the release of FSH. The key word in this question is "releasing". The organ that produces releasing or inhibiting factors is the hypothalamus. (A): The anterior pituitary is the master gland that secretes several hormones. (B): The posterior pituitary secretes oxytocin and vasopressin. (D): The pineal gland is a gland at the base of the brain that secretes melatonin and helps regulate circadian rhythms. (E): The ovary secretes estrogen and progesterone.

34. **(B) is the correct answer.** Eutrophication is the aging process through which a body of water becomes choked by an overabundance of green algae. This causes lakes to become marshes, and then terrestrial communities. (A): Succession is the gradual progression of one community into another. (C): Evolution is the gradual change in a population over time. (D): The greenhouse effect is the gradual warming of the earth's surface by increasing concentrations of CO_2 in the atmosphere. (E): Lake turnover is the vertical movement of water brought on by seasonal temperature changes.

35. **(C) is the correct answer.** The principle inorganic compound found in living things is water. Water is a necessary component for life.

36. **(C) is the correct answer.** The loop of Henle is responsible for the concentration of urine. The longer the loop of Henle, the more water would be reabsorbed which would make the filtrate more concentrated. (D): If the loop of Henle were longer, the collecting ducts may or may not also be longer. (A): If the walls of the nephrons were thicker and impermeable, water would not be conserved. (E): The proximal convoluted tubules are not associated with the concentration of urine.

37. **(E) is the correct answer.** All of the following are modes of asexual reproduction except meiosis, which is the production of gametes for sexual reproduction. (A): Sporulation is a form of asexual reproduction in which spores are produced. (B): Fission is the equal division of a bacterial cell. (C): Budding is a form of asexual reproduction in yeasts in which small cells grow from a parent cell. (D): Cloning refers to a population of cells descended by mitotic division from one cell.

38. **(B) is the correct answer.** The moist skin of earthworms, the lenticels of plants, and the spiracles of grasshoppers are all associated with the process of respiration—gas exchange.

39. **(B) is the correct answer.** Locomotion in annelids is accomplished through the interaction of muscles and paired setae. Setae are bristles that help earthworms grip the soil as they inch forward. (A), (C), and (D): Insects have an exoskeleton and jointed appendages to help them move. (E): Amoeba have pseudopods to help them move.

40. **(B) is the correct answer.** All of the following are examples of connective tissue except muscle. Connective tissue connects and supports other tissues. Ligaments, blood, cartilage, and bone are all connective tissues.

41. **(D) is the correct answer.** In most plants, germination is triggered by the presence of oxygen, water, and temperature. Use Process of Elimination for this question. Soil is not necessary for germination; eliminate answer choice (A). All plants require water so eliminate answer choice (E). All plants require oxygen; eliminate answer choices (B) and (C).

42. **(E) is the correct answer.** Each smaller trophic level has less available energy than the previous level. Use Process of Elimination for this question. (A): A net *loss* occurs as energy is transferred from one organism to another. (B): The total energy in plants is *more* than that in herbivores. (C): The first tropic level is at the *bottom* of a pyramid. (D): The total mass of carnivores is *less* than the total mass of plants.

43. **(B) is the correct answer.** When a forest of trees is devastated by fire, secondary succession is most likely to occur. Secondary succession refers to ecological succession in a disturbed community. (A): Plants and animals will continue to inhabit the region once the community is reestablished. (C): Tough grasses are not pioneers. (D): It cannot be determined if the number of species will be stabilized. (E): The forest could support living things once new plants took root and grew.

44. **(B) is the correct answer.** Translation, the synthesis of proteins from mRNA, occurs in the cytoplasm. (I): DNA replication occurs in the nucleus. (II): Transcription, the synthesis of RNA from DNA, occurs in the nucleus.

45. **(C) is the correct answer.** Crossing over permits scientists to determine chromosome mapping. Chromosome mapping is a detailed map of all the genes on a chromosome. The frequency of crossing-over between any two alleles is proportional to the distance between them. The further apart the two linked alleles are on a chromosome, the more often the chromosome will break between them. Crossing-over does not tell us about the chance of variation in zygotes, the rate of mutations, or whether the traits are dominant, recessive, or masked.

46. **(E) is the correct answer.** Legume plants assimilate nitrogen through the activity of small microorganisms. Review the nitrogen cycle. Legumes are able to take in nitrogen and convert it to nitrates with the help of nitrogen-fixing bacteria. (A): Transpiration in plants is not part of the nitrogen cycle. (B): Green plants do not take in nitrogen in the form of ammonia. (C): Soil bacteria convert atmospheric nitrogen into nitrates.

47. **(E) is the correct answer.** The results of the process of cloning are most similar to the results of the process of mitosis. Cloning is the asexual reproduction of genetically identical cells or organisms. The only answer choice that is a form of asexual reproduction is mitosis—new cells produced from a parent cell. (A): Gametogenesis is a series of cell divisions that leads to the production of gametes. (B): Fertilization is the union of egg and sperm to form a zygote. (C): Pollination is the process of transferring pollen grains from the anther to the stigma of a flower. (D): Meiosis is the process of cell division that produces gametes (non-identical cells).

48. **(B) is the correct answer.** The most likely explanation for this phenomenon is that these birds have different ecological niches. An ecological niche is the position or function of an organism or population in its environment. (A): We do not know if there is a short supply of resources. (C): We do not know how long the bird species live together. (D): The breeding patterns of the bird species does not explain the lack of competition. (E): A habitat is the physical place or environment in which an organism normally lives. These birds clearly share the same habitat.

49. **(B) is the correct answer.** This relationship is an example of parasitism. Parasitism is a form of symbiosis in which one organism benefits and the other is harmed. (A): Commensalism is a form of symbiosis in which one organism benefits and the other is unaffected. (C): Mutualism is a form of symbiosis in which both organisms benefit. (D): Gravitropism is the growth of a plant toward or away from gravity. (E): Thigmotropism refers to growth stimulated by contact with an object.

50. **(C) is the correct answer.** If the nucleotide sequence of a DNA molecule is 5′-C-A-T-3′, then the transcribed DNA strand (mRNA) would be 3′-G-U-A-5′. The nucleotide sequence of the tRNA codon would be 5′-C-A-U-3′.

51. **(A) is the correct answer.** Viruses are considered an exception to the cell theory because they are not independent organisms. They can only survive by invading a host. (B): Viruses have either DNA or RNA. (C): Not all viruses have a tail. (D): Viruses did not evolve from ancestral protists. (E): Viruses have no nuclei.

52. **(C) is the correct answer.** All of the following organs in the digestive system secrete digestive enzymes except the gall bladder. The gall bladder stores and secretes bile produced by the liver, which is an emulsifier, not an enzyme. (A): The mouth secretes salivary amylase. (B): The stomach secretes pepsin. (D): The small intestine secretes protease. (E): The pancreas secretes pancreatic amylase and lipase.

53. **(C) is the correct answer.** Memory loss would most likely be due to a malfunction of the cerebrum. The cerebrum controls all voluntary activities and receives and interprets sensory information. (A): The medulla controls involuntary actions such as breathing. (B): The cerebellum coordinates muscle activity and controls balance. (D): The pons is a mass of nerve fibers running across the surface of the mammalian brain. (E): The hypothalamus maintains homeostasis.

54. **(D) is the correct answer.** In nonplacental mammals, the embryo obtains its food from the yolk sac. The yolk sac provides food for the embryo. (A): The ovary is where eggs mature. (B): The uterus is the organ that contains the developing embryo. (C): The oviduct (also known as the fallopian tube) is a tube that carries the egg from the ovary toward the uterus. (E): The allantois is an extraembryonic sac that gets rid of wastes.

55. **(B) is the correct answer.** The similarity suggests that humans and chimpanzees are more closely related than humans and dogs. Since these two organisms share similar amino acid sequences, they must share more recent common ancestors than with the dog.

56. **(B) is the correct answer.** The event that had to occur before oxygen filled the atmosphere was that autotrophs, which make their own food and give off oxygen, had to evolve.

57. **(E) is the correct answer.** The structure in an earthworm that has a function similar to that of the alveoli of a human is the skin, which is where gas exchange takes place in an earthworm, and from which moisture is lost. (A): The excretory organs in insects are called malpighian tubules. (C): The chitinous exoskeleton is the protective layer of insects. (D): Gills are the respiratory organs in aquatic animals. (B) Nephridia are the excretory organs of invertebrates.

58. **(E) is the correct answer.** Parathyroid hormone raises calcium and lowers phosphate levels in the blood. Calcium phosphate is a component of bone.

59. **(A) is the correct answer.** Cortisol is a hormone (produced by the adrenal cortex) that increases blood sugar concentration.

60. **(C) is the correct answer.** Progesterone is a sex hormone responsible for preparing the body for pregnancy.

61. **(D) is the correct answer.** The hypothalamus secretes hormones that stimulate or inhibit the actions of the anterior pituitary.

62. **(D) is the correct answer.** Classical conditioning—also known as Pavlovian conditioning—is when a certain stimulus is repeatedly coupled with a stimulus triggering a behavioral response.

63. **(C) is the correct answer.** Imprinting is the brief period during which ducklings associate any moving object with their mother.

64. **(B) is the correct answer.** Operant conditioning is learning based on trial and error.

65. **(E) is the correct answer.** A circadian rhythm is an internal cycle (biological clock) that repeats approximately every 24 hours. Plants placed in continued darkness continue to open and close their stomates on a 24-hour cycle.

66. **(B) is the correct answer.** The ectoderm gives rise to the skin and nervous system.

67. **(C) is the correct answer.** The endoderm gives rise to the inner linings of the digestive tract and associated organs.

68. **(A) is the correct answer.** The mesoderm gives rise to the skeletal, muscular, and connective tissues.

69. **(E) is the correct answer.** The tropical rain forest is the biome that has soil that's depleted of nutrients.

70. **(B) is the correct answer.** The tundra is a biome that's characterized by a short growing season because it's seasonally covered by snow.

71. **(A) is the correct answer.** Taiga is a biome characterized by black bears, moose, and wolves.

72. **(C) is the correct answer.** The embryo develops in the ovary of the plant.

73. **(D) is the correct answer.** The anther has a similar function to that of the human testes, in that they both produce the male gametes.

74. **(A) is the correct answer.** The stigma is the sticky portion of the flowering plant that traps pollen grains.

75. **(D) is the correct answer.** Meiosis in flowering plants occurs within the ovary and the anthers—microspores are produced by the male reproductive organs, the anthers.

76. **(E) is the correct answer.** A nucleotide is a building block of nucleic acids.

77. **(C) is the correct answer.** Fatty acids are a component of phospholipids.

78. **(A) is the correct answer.** Monosaccharides (like glucose) are a major energy source for the body. (C) is also an energy source but not a major one.

79. **(D) is the correct answer.** Amino acids are the building blocks of proteins.

80. **(D) is the correct answer.** The graph shows the average number of offspring per female per day. Since there are 100 females in the $\frac{1}{2}$ liter, the total number of offspring on the twentieth day would be 100 females times 2 offspring per day, which equals 200 offspring.

81. **(D) is the correct answer.** The fact that lines A and B decrease suggests that the reproductive rate is related to the population density. After several weeks, as the number of offspring in both jars increase, the reproductive rate decreases.

82. **(B) is the correct answer.** Based on the graph the number of *Daphnia* offspring in jar B is half that of jar A and the reproductive rate is half. Therefore, the reproductive rate in jar B is the the same as the reproductive rate in jar A.

83. **(B) is the correct answer.** The largest amount of energy is available to producers. Population B is most likely composed of producers, since they have the largest biomass.

84. **(B) is the correct answer.** An increase in the number of organisms in population C would most likely lead to a decrease in the biomass of B since population B is the food source for population C. Make a pyramid based on the biomasses given. If population C increases, population B decreases. (A) and (C): We cannot necessarily predict what will happen to the biomass of populations that are above population C. (D): The food source available to population C would most likely decrease, not increase. (E): We do not know if an increase in the number of organisms will lead to intense competition for the food resources. It depends on how large the population increase is. If the population increase is small, the organisms would probably not have to compete for food.

85. **(C) is the correct answer.** A 90 percent reduction of productivity in the grass is a reduction of 9,000 pounds. That means the grass should be able to support 1,000 pounds of crickets.

86. **(B) is the correct answer.** Chromosomes replicate during interphase, the S phase. (A) and (C): During G_1 and G_2, the cell makes protein and performs other metabolic duties. (E): Cytokinesis refers to cytoplasmic division of the cell.

87. **(A) is the correct answer.** One of the first signs of prophase in mammalian cells is the appearance of chromosomes.

88. **(C) is the correct answer.** Mitosis occurs in all of the following type of cells except mature white blood cells. Mature white blood cells are short-lived and do not divide.

89. **(B) is the correct answer.** Since neurons are not capable of dividing, it is reasonable to conclude that these cells will not complete the G_1 phase. This is a reading comprehension question. The passage states that cells that do not divide are arrested at the G_1 phase. (A): These cells will not be committed to go through cell division. (C) and (D): These cells will not enter the G_2 or S-phase. (E): The duration of the cell cycle would be brief, not long.

90. **(A) is the correct answer.** According to the graph, the resting membrane potential of the muscle fiber is close to –90mV.

91. **(E) is the correct answer.** Refer to the second graph about the membrane permeability of ions during a muscle contraction. During depolarization, the membrane is permeable to Na^+. (A): The voltage-gated K^+ channels do not open until after depolarization. (B): The concentration of Ca^{2+} does not become more negative. (C): The membrane potential changes from –90 mV to +20 mV.

92. **(D) is the correct answer.** Refer to both graphs in the passage. In cardiac fibers, the duration of an action potential—a neuronal impulse—is approximately 0.3 seconds.

93. **(C) is the correct answer.** In cardiac muscle fibers, depolarization is prolonged compared to that in skeletal muscle fibers. (A): The membrane is permeable to both Na^+ and K^+. (B): In cardiac muscle fibers, voltage-gated K^+ channels open during repolarization. (D): The refractory period is longer in cardiac muscle fibers compared to skeletal muscle fibers. (E): The speed of a contraction is faster in skeletal muscle fibers.

94. **(C) is the correct answer.** The body plan associated with nematodes is Organism 3. Nematodes—roundworms—are pseudocoelomates. They have a body cavity, but the cavity is only partially lined with mesodermally-derived tissue.

95. **(B) is the correct answer.** The body plan associated with flatworms is organism 2. Flatworms have no body cavity. They are acoelomates.

96. **(E) is the correct answer.** Organism 5 should be placed in the phylum chordata. Coelomates have a body cavity that forms distinct cell layers (ectoderm, mesoderm, and endoderm). The only answer choice with complex processes: coelmates deuterostomes, and bilateral symmetry is chordata.

97. **(A) is the correct answer.** All of the following organisms exhibit bilateral symmetry except hydra. Organisms with bilateral symmetry have two similar halves on either side of a central plane. Animals with radial symmetry have body parts arranged around a central line. Animals that exhibit radial symmetry are hydra, jellyfish, and sea anemones.

98. **(B) is the correct answer.** When the level of CTP is low in a cell, the metabolic traffic down the pathway increases. This pathway is controlled by feedback inhibition. The final product of the pathway inhibits the activity of the first enzyme. When the supply of CTP is low, the pathway will continue to produce CTP.

99. **(B) is the correct answer.** This enzymatic phenomenon is an example of feedback inhibition. Feedback inhibition is the metabolic regulation in which high levels of an enzymatic pathway's final product inhibit the activity of its rate-limiting enzyme. (A): Transcription is the production of RNA from DNA. (C): Dehydration synthesis is the formation of a covalent bond by the removal of water. (D): In photosynthesis, radiant energy is converted to chemical energy. (E): Hydrolysis is the breaking of a covalent bond by adding water.

100. **(C) is the correct answer.** The biosynthesis of cytidine 5′-triphate requires a nitrogenous base, three phosphates, and the sugar ribose. Pyrimidines are a class of nitrogenous bases with a single ring structure. The sugar they contain is ribose (shown in the pathway).

ESSAY 1

Chlorophyll is the principle light-harnessing pigment found in the thylakoid membrane of chloroplasts. It is a molecule that's made up of a large ring structure composed of smaller rings (pyrroles). In the center of the large porphyrin ring is a magnesium atom surrounded by four nitrogen atoms. The small rings consist of many alternating single and double bonds. These alternating bonds allow electrons in the porphyrin ring and magnesium atoms to move around freely. Alternating double and single bonds are commonly found in molecules that strongly absorb visible light. Photosynthesis occurs when sunlight activates chlorophyll by exciting their electrons. The chlorophyll molecule has a long hydrophobic hydrocarbon tail. This shape allows many chlorophyll molecules to be grouped together like a stack of saucers.

The following experiment could determine the influence of sunlight on chlorophyll. First, extract chlorophyll from the leaf by boiling it gently in a dissolved solution. Then pass light of a known wavelength through the solution and measure it, using a spectrophotometer. The wavelength of the entering light can be varied to see which wavelength is most absorbed by the solution. You could then plot the data on a graph to form an absorption spectrum. The absorption spectrum will show two peaks, which represent the type of light absorbed by the pigment. The valley would represent the light that's reflected.

Based on this experiment, it would be inferred that certain wavelengths, such as orange-red and violet-blue, are strongly absorbed by chlorophyll, whereas green and yellow hues are least absorbed, and are reflected. Chlorophyll gives plants their characteristic green color because it reflects yellow-green light.

ESSAY 2

Hormones are chemical messengers that travel through the blood and act on target cells in the body. Hormone secretion is regulated by a negative feedback system, which is the basis of hormonal regulation. In part, negative feedback loops mean that an excess of a hormone in the bloodstream temporarily shuts down the production of that hormone.

The anterior pituitary gland secretes a hormone called thyroid-stimulating hormone (TSH). TSH stimulates the thyroid gland to secrete its hormone; thyroxine. Thyroxine regulates the basal metabolic rate in most body tissues. If the blood level of thyroxine rises above normal, it will suppress the secretion of TSH which, in turn, will lower the blood level of thyroxine. Both responses are negative feedback mechanisms.

Two hormones play a critical role in the regulation of calcium: parathyroid hormone and calcitonin. Parathyroid hormone, which is secreted from the parathyroid glands, increases blood calcium levels. It triggers the bones to release the calcium stored inside them. Calcitonin acts as an antagonist of parathyroid hormone. It decreases blood calcium levels. When blood calcium levels are above normal, the thyroid gland secretes calcitonin. This mechanism helps to maintain homeostasis.

ACTH is another hormone that's released by the anterior pituitary. It stimulates the adrenal cortex to secrete a number of steroid hormones, including cortisol. Cortisol increases the blood's concentration of glucose and helps the body adapt to stress. When cortisol levels reach a peak level in the bloodstream, the anterior pituitary is temporarily prevented from producing ACTH. This, in turn, shuts down the production of cortisol. Once the level of cortisol falls below normal, the adrenal cortex resumes production of cortisol.

Essay 3

A gene is a heredity unit located at a specific locus along a chromosome. Genes are made up of DNA, and DNA is made up of repeating subunits of nucleotides. A nucleotide consists of three parts: a five-carbon sugar, a phosphate group, and a nitrogenous base.

When a chromosome replicates, the two DNA strands unwind and the hydrogen bonds between them are broken. Each strand serves as a template for the synthesis of a complementary strand. Once the process is initiated (with an RNA primer), DNA polymerase adds nucleotides to each growing strand. One strand serves as the leading strand (it is made continuously) and the other serves as the lagging strand (it is made discontinuously). Each base matches the appropriate bases in the template strand; they are complementary. Once the complementary strands are formed, hydrogen bonds form between the new base pairs, leaving two identical copies of the original DNA molecule.

A gene mutation is a change in the sequence of base pairs in a DNA. It results from defects in the sequence of bases. Mutations that involve a single base change in the DNA sequence are called point mutations. These are base substitutions involving a single DNA nucleotide being replaced by another nucleotide. Another type of mutation is a frameshift mutation, caused by an insertion or deletion of bases in the DNA sequence. This causes a shift in the reading frame of DNA.

Essay 4

Since the amount of intracellular water is critical, one major challenge that various organisms faced was how they would regulate body water and get rid of wastes. Waste products include carbon dioxide, salts, and nitrogenous wastes. This issue is especially important because nitrogenous wastes are highly toxic to the body. Over the course of evolution, organisms developed various structural adaptations to deal with this problem.

Insects developed excretory organs called Malphigian tubules. These long, slender tubules take up water and salt, concentrate the waste, and empty it into the intestine. The waste product, uric acid, is excreted as a dry pellet through the anus. Insects also have a hard, dry cuticle that has a waxy outer layer; this aids in preventing water loss.

Organisms that live in the sea, such as marine fishes, had to cope with the gradual increase in the saltiness of the water. Seawater is hypertonic relative to their body fluids, and marine fish had to protect their body cells from water loss because without a specific mechanism for this, they would become dehydrated even though they're surrounded by water. They excrete concentrated urine that is isotonic with their body fluids. They also eliminate excess salt by actively pumping it out through their gills.

Since humans are terrestrial organisms, they must conserve plenty of water. Humans also must get rid of nitrogenous wastes and their major excretory organ is the kidney. Each kidney is made up of millions of function units called nephrons. As the filtrate travels along the nephron, glucose, amino acids, and salts are retained by the body and the rest of the fluid is concentrated into urine. The skin is another important organ that gets rid of wastes. It contains sweat glands that help to maintain an optimal salt balance in the body.

Index

ABOUT THE AUTHOR

Kim Magloire received a degree in Biology from Princeton University. She is now completing her doctorate in Epidemiology at Columbia University. As a teacher for The Princeton Review, Kim has prepared hundreds of students for the AP Biology Test.

Kim Magloire is also founder of SciTech Youth Foundation, a non-profit organization that inspires interest in the sciences by hosting a variety of events for high school students. She is also the publisher of SciTech Magazine, a high school publication that produces a wide-range of articles on science, math, and technology. Kim also gives presentations at the New York Public Library on science-related topics.

The Princeton Review

Completely darken bubbles with a No. 2 pencil. If you make a mistake, be sure to erase mark completely. Erase all stray marks.

1. YOUR NAME: _____
(Print)

| Last | First | M.I. |

SIGNATURE: _____   **DATE:** _____ / _____ / _____

HOME ADDRESS: _____
(Print)

Number and Street

_____
City   State   Zip Code

PHONE NO. : _____
(Print)

IMPORTANT: Please fill in these boxes exactly as shown on the back cover of your test book.

2. TEST FORM

_____

6. DATE OF BIRTH

Month		Day		Year	
○ JAN					
○ FEB					
○ MAR	⓪	⓪	⓪	⓪	
○ APR	①	①	①	①	
○ MAY	②	②	②	②	
○ JUN	③	③	③	③	
○ JUL		④	④	④	
○ AUG		⑤	⑤	⑤	
○ SEP		⑦	⑦	⑦	
○ OCT		⑧	⑧	⑧	
○ NOV		⑨	⑨	⑨	
○ DEC					

3. TEST CODE   **4. REGISTRATION NUMBER**

⓪	Ⓐ	⓪	⓪	⓪	⓪	⓪	⓪	⓪	⓪	⓪
①	Ⓑ	①	①	①	①	①	①	①	①	①
②	Ⓒ	②	②	②	②	②	②	②	②	②
③	Ⓓ	③	③	③	③	③	③	③	③	③
④	Ⓔ	④	④	④	④	④	④	④	④	④
⑤	Ⓕ	⑤	⑤	⑤	⑤	⑤	⑤	⑤	⑤	⑤
⑦	Ⓖ	⑦	⑦	⑦	⑦	⑦	⑦	⑦	⑦	⑦
⑧		⑧	⑧	⑧	⑧	⑧	⑧	⑧	⑧	⑧
⑨		⑨	⑨	⑨	⑨	⑨	⑨	⑨	⑨	⑨

7. SEX

○ MALE
○ FEMALE

The Princeton Review
© 1996 Princeton Review L.L.C.
FORM NO. 00001-PR

5. YOUR NAME

First 4 letters of last name				FIRST INIT	MID INIT
Ⓐ	Ⓐ	Ⓐ	Ⓐ	Ⓐ	Ⓐ
Ⓑ	Ⓑ	Ⓑ	Ⓑ	Ⓑ	Ⓑ
Ⓒ	Ⓒ	Ⓒ	Ⓒ	Ⓒ	Ⓒ
Ⓓ	Ⓓ	Ⓓ	Ⓓ	Ⓓ	Ⓓ
Ⓔ	Ⓔ	Ⓔ	Ⓔ	Ⓔ	Ⓔ
Ⓕ	Ⓕ	Ⓕ	Ⓕ	Ⓕ	Ⓕ
Ⓖ	Ⓖ	Ⓖ	Ⓖ	Ⓖ	Ⓖ
Ⓗ	Ⓗ	Ⓗ	Ⓗ	Ⓗ	Ⓗ
Ⓘ	Ⓘ	Ⓘ	Ⓘ	Ⓘ	Ⓘ
Ⓙ	Ⓙ	Ⓙ	Ⓙ	Ⓙ	Ⓙ
Ⓚ	Ⓚ	Ⓚ	Ⓚ	Ⓚ	Ⓚ
Ⓛ	Ⓛ	Ⓛ	Ⓛ	Ⓛ	Ⓛ
Ⓜ	Ⓜ	Ⓜ	Ⓜ	Ⓜ	Ⓜ
Ⓝ	Ⓝ	Ⓝ	Ⓝ	Ⓝ	Ⓝ
Ⓞ	Ⓞ	Ⓞ	Ⓞ	Ⓞ	Ⓞ
Ⓟ	Ⓟ	Ⓟ	Ⓟ	Ⓟ	Ⓟ
Ⓠ	Ⓠ	Ⓠ	Ⓠ	Ⓠ	Ⓠ
Ⓡ	Ⓡ	Ⓡ	Ⓡ	Ⓡ	Ⓡ
Ⓢ	Ⓢ	Ⓢ	Ⓢ	Ⓢ	Ⓢ
Ⓣ	Ⓣ	Ⓣ	Ⓣ	Ⓣ	Ⓣ
Ⓤ	Ⓤ	Ⓤ	Ⓤ	Ⓤ	Ⓤ
Ⓥ	Ⓥ	Ⓥ	Ⓥ	Ⓥ	Ⓥ
Ⓦ	Ⓦ	Ⓦ	Ⓦ	Ⓦ	Ⓦ
Ⓧ	Ⓧ	Ⓧ	Ⓧ	Ⓧ	Ⓧ
Ⓨ	Ⓨ	Ⓨ	Ⓨ	Ⓨ	Ⓨ
Ⓩ	Ⓩ	Ⓩ	Ⓩ	Ⓩ	Ⓩ

Section ①

Start with number 1 for each new section.
If a section has fewer questions than answer spaces, leave the extra answer spaces blank.

1. Ⓐ Ⓑ Ⓒ Ⓓ Ⓔ
2. Ⓐ Ⓑ Ⓒ Ⓓ Ⓔ
3. Ⓐ Ⓑ Ⓒ Ⓓ Ⓔ
4. Ⓐ Ⓑ Ⓒ Ⓓ Ⓔ
5. Ⓐ Ⓑ Ⓒ Ⓓ Ⓔ
6. Ⓐ Ⓑ Ⓒ Ⓓ Ⓔ
7. Ⓐ Ⓑ Ⓒ Ⓓ Ⓔ
8. Ⓐ Ⓑ Ⓒ Ⓓ Ⓔ
9. Ⓐ Ⓑ Ⓒ Ⓓ Ⓔ
10. Ⓐ Ⓑ Ⓒ Ⓓ Ⓔ
11. Ⓐ Ⓑ Ⓒ Ⓓ Ⓔ
12. Ⓐ Ⓑ Ⓒ Ⓓ Ⓔ
13. Ⓐ Ⓑ Ⓒ Ⓓ Ⓔ
14. Ⓐ Ⓑ Ⓒ Ⓓ Ⓔ
15. Ⓐ Ⓑ Ⓒ Ⓓ Ⓔ
16. Ⓐ Ⓑ Ⓒ Ⓓ Ⓔ
17. Ⓐ Ⓑ Ⓒ Ⓓ Ⓔ
18. Ⓐ Ⓑ Ⓒ Ⓓ Ⓔ
19. Ⓐ Ⓑ Ⓒ Ⓓ Ⓔ
20. Ⓐ Ⓑ Ⓒ Ⓓ Ⓔ
21. Ⓐ Ⓑ Ⓒ Ⓓ Ⓔ
22. Ⓐ Ⓑ Ⓒ Ⓓ Ⓔ
23. Ⓐ Ⓑ Ⓒ Ⓓ Ⓔ
24. Ⓐ Ⓑ Ⓒ Ⓓ Ⓔ
25. Ⓐ Ⓑ Ⓒ Ⓓ Ⓔ
26. Ⓐ Ⓑ Ⓒ Ⓓ Ⓔ
27. Ⓐ Ⓑ Ⓒ Ⓓ Ⓔ
28. Ⓐ Ⓑ Ⓒ Ⓓ Ⓔ
29. Ⓐ Ⓑ Ⓒ Ⓓ Ⓔ
30. Ⓐ Ⓑ Ⓒ Ⓓ Ⓔ

31. Ⓐ Ⓑ Ⓒ Ⓓ Ⓔ
32. Ⓐ Ⓑ Ⓒ Ⓓ Ⓔ
33. Ⓐ Ⓑ Ⓒ Ⓓ Ⓔ
34. Ⓐ Ⓑ Ⓒ Ⓓ Ⓔ
35. Ⓐ Ⓑ Ⓒ Ⓓ Ⓔ
36. Ⓐ Ⓑ Ⓒ Ⓓ Ⓔ
37. Ⓐ Ⓑ Ⓒ Ⓓ Ⓔ
38. Ⓐ Ⓑ Ⓒ Ⓓ Ⓔ
39. Ⓐ Ⓑ Ⓒ Ⓓ Ⓔ
40. Ⓐ Ⓑ Ⓒ Ⓓ Ⓔ
41. Ⓐ Ⓑ Ⓒ Ⓓ Ⓔ
42. Ⓐ Ⓑ Ⓒ Ⓓ Ⓔ
43. Ⓐ Ⓑ Ⓒ Ⓓ Ⓔ
44. Ⓐ Ⓑ Ⓒ Ⓓ Ⓔ
45. Ⓐ Ⓑ Ⓒ Ⓓ Ⓔ
46. Ⓐ Ⓑ Ⓒ Ⓓ Ⓔ
47. Ⓐ Ⓑ Ⓒ Ⓓ Ⓔ
48. Ⓐ Ⓑ Ⓒ Ⓓ Ⓔ
49. Ⓐ Ⓑ Ⓒ Ⓓ Ⓔ
50. Ⓐ Ⓑ Ⓒ Ⓓ Ⓔ
51. Ⓐ Ⓑ Ⓒ Ⓓ Ⓔ
52. Ⓐ Ⓑ Ⓒ Ⓓ Ⓔ
53. Ⓐ Ⓑ Ⓒ Ⓓ Ⓔ
54. Ⓐ Ⓑ Ⓒ Ⓓ Ⓔ
55. Ⓐ Ⓑ Ⓒ Ⓓ Ⓔ
56. Ⓐ Ⓑ Ⓒ Ⓓ Ⓔ
57. Ⓐ Ⓑ Ⓒ Ⓓ Ⓔ
58. Ⓐ Ⓑ Ⓒ Ⓓ Ⓔ
59. Ⓐ Ⓑ Ⓒ Ⓓ Ⓔ
60. Ⓐ Ⓑ Ⓒ Ⓓ Ⓔ

61. Ⓐ Ⓑ Ⓒ Ⓓ Ⓔ
62. Ⓐ Ⓑ Ⓒ Ⓓ Ⓔ
63. Ⓐ Ⓑ Ⓒ Ⓓ Ⓔ
64. Ⓐ Ⓑ Ⓒ Ⓓ Ⓔ
65. Ⓐ Ⓑ Ⓒ Ⓓ Ⓔ
66. Ⓐ Ⓑ Ⓒ Ⓓ Ⓔ
67. Ⓐ Ⓑ Ⓒ Ⓓ Ⓔ
68. Ⓐ Ⓑ Ⓒ Ⓓ Ⓔ
69. Ⓐ Ⓑ Ⓒ Ⓓ Ⓔ
70. Ⓐ Ⓑ Ⓒ Ⓓ Ⓔ
71. Ⓐ Ⓑ Ⓒ Ⓓ Ⓔ
72. Ⓐ Ⓑ Ⓒ Ⓓ Ⓔ
73. Ⓐ Ⓑ Ⓒ Ⓓ Ⓔ
74. Ⓐ Ⓑ Ⓒ Ⓓ Ⓔ
75. Ⓐ Ⓑ Ⓒ Ⓓ Ⓔ
76. Ⓐ Ⓑ Ⓒ Ⓓ Ⓔ
77. Ⓐ Ⓑ Ⓒ Ⓓ Ⓔ
78. Ⓐ Ⓑ Ⓒ Ⓓ Ⓔ
79. Ⓐ Ⓑ Ⓒ Ⓓ Ⓔ
80. Ⓐ Ⓑ Ⓒ Ⓓ Ⓔ
81. Ⓐ Ⓑ Ⓒ Ⓓ Ⓔ
82. Ⓐ Ⓑ Ⓒ Ⓓ Ⓔ
83. Ⓐ Ⓑ Ⓒ Ⓓ Ⓔ
84. Ⓐ Ⓑ Ⓒ Ⓓ Ⓔ
85. Ⓐ Ⓑ Ⓒ Ⓓ Ⓔ
86. Ⓐ Ⓑ Ⓒ Ⓓ Ⓔ
87. Ⓐ Ⓑ Ⓒ Ⓓ Ⓔ
88. Ⓐ Ⓑ Ⓒ Ⓓ Ⓔ
89. Ⓐ Ⓑ Ⓒ Ⓓ Ⓔ
90. Ⓐ Ⓑ Ⓒ Ⓓ Ⓔ

91. Ⓐ Ⓑ Ⓒ Ⓓ Ⓔ
92. Ⓐ Ⓑ Ⓒ Ⓓ Ⓔ
93. Ⓐ Ⓑ Ⓒ Ⓓ Ⓔ
94. Ⓐ Ⓑ Ⓒ Ⓓ Ⓔ
95. Ⓐ Ⓑ Ⓒ Ⓓ Ⓔ
96. Ⓐ Ⓑ Ⓒ Ⓓ Ⓔ
97. Ⓐ Ⓑ Ⓒ Ⓓ Ⓔ
98. Ⓐ Ⓑ Ⓒ Ⓓ Ⓔ
99. Ⓐ Ⓑ Ⓒ Ⓓ Ⓔ
100. Ⓐ Ⓑ Ⓒ Ⓓ Ⓔ
101. Ⓐ Ⓑ Ⓒ Ⓓ Ⓔ
102. Ⓐ Ⓑ Ⓒ Ⓓ Ⓔ
103. Ⓐ Ⓑ Ⓒ Ⓓ Ⓔ
104. Ⓐ Ⓑ Ⓒ Ⓓ Ⓔ
105. Ⓐ Ⓑ Ⓒ Ⓓ Ⓔ
106. Ⓐ Ⓑ Ⓒ Ⓓ Ⓔ
107. Ⓐ Ⓑ Ⓒ Ⓓ Ⓔ
108. Ⓐ Ⓑ Ⓒ Ⓓ Ⓔ
109. Ⓐ Ⓑ Ⓒ Ⓓ Ⓔ
110. Ⓐ Ⓑ Ⓒ Ⓓ Ⓔ
111. Ⓐ Ⓑ Ⓒ Ⓓ Ⓔ
112. Ⓐ Ⓑ Ⓒ Ⓓ Ⓔ
113. Ⓐ Ⓑ Ⓒ Ⓓ Ⓔ
114. Ⓐ Ⓑ Ⓒ Ⓓ Ⓔ
115. Ⓐ Ⓑ Ⓒ Ⓓ Ⓔ
116. Ⓐ Ⓑ Ⓒ Ⓓ Ⓔ
117. Ⓐ Ⓑ Ⓒ Ⓓ Ⓔ
118. Ⓐ Ⓑ Ⓒ Ⓓ Ⓔ
119. Ⓐ Ⓑ Ⓒ Ⓓ Ⓔ
120. Ⓐ Ⓑ Ⓒ Ⓓ Ⓔ

Completely darken bubbles with a No. 2 pencil. If you make a mistake, be sure to erase mark completely. Erase all stray marks.

1. YOUR NAME: _____
(Print)   Last   First   M.I.

SIGNATURE: _____   **DATE:** _____ / _____ / _____

HOME ADDRESS: _____
(Print)   Number and Street

_____
City   State   Zip Code

PHONE NO. : _____
(Print)

IMPORTANT: Please fill in these boxes exactly as shown on the back cover of your test book.

2. TEST FORM

3. TEST CODE   **4. REGISTRATION NUMBER**

⓪	Ⓐ	⓪	⓪	⓪	⓪	⓪	⓪	⓪	⓪	⓪
①	Ⓑ	①	①	①	①	①	①	①	①	①
②	Ⓒ	②	②	②	②	②	②	②	②	②
③	Ⓓ	③	③	③	③	③	③	③	③	③
④	Ⓔ	④	④	④	④	④	④	④	④	④
⑤	Ⓕ	⑤	⑤	⑤	⑤	⑤	⑤	⑤	⑤	⑤
⑦	Ⓖ	⑦	⑦	⑦	⑦	⑦	⑦	⑦	⑦	⑦
⑧		⑧	⑧	⑧	⑧	⑧	⑧	⑧	⑧	⑧
⑨		⑨	⑨	⑨	⑨	⑨	⑨	⑨	⑨	⑨

6. DATE OF BIRTH

Month		Day		Year	
◯ JAN					
◯ FEB					
◯ MAR	⓪	⓪	⓪	⓪	
◯ APR	①	①	①	①	
◯ MAY	②	②	②	②	
◯ JUN	③	③	③	③	
◯ JUL		④	④	④	
◯ AUG		⑤	⑤	⑤	
◯ SEP		⑦	⑦	⑦	
◯ OCT		⑧	⑧	⑧	
◯ NOV		⑨	⑨	⑨	
◯ DEC					

7. SEX
◯ MALE
◯ FEMALE

The Princeton Review
© 1996 Princeton Review L.L.C.
FORM NO. 00001-PR

5. YOUR NAME

First 4 letters of last name				FIRST INIT	MID INIT
Ⓐ	Ⓐ	Ⓐ	Ⓐ	Ⓐ	Ⓐ
Ⓑ	Ⓑ	Ⓑ	Ⓑ	Ⓑ	Ⓑ
Ⓒ	Ⓒ	Ⓒ	Ⓒ	Ⓒ	Ⓒ
Ⓓ	Ⓓ	Ⓓ	Ⓓ	Ⓓ	Ⓓ
Ⓔ	Ⓔ	Ⓔ	Ⓔ	Ⓔ	Ⓔ
Ⓕ	Ⓕ	Ⓕ	Ⓕ	Ⓕ	Ⓕ
Ⓖ	Ⓖ	Ⓖ	Ⓖ	Ⓖ	Ⓖ
Ⓗ	Ⓗ	Ⓗ	Ⓗ	Ⓗ	Ⓗ
Ⓘ	Ⓘ	Ⓘ	Ⓘ	Ⓘ	Ⓘ
Ⓙ	Ⓙ	Ⓙ	Ⓙ	Ⓙ	Ⓙ
Ⓚ	Ⓚ	Ⓚ	Ⓚ	Ⓚ	Ⓚ
Ⓛ	Ⓛ	Ⓛ	Ⓛ	Ⓛ	Ⓛ
Ⓜ	Ⓜ	Ⓜ	Ⓜ	Ⓜ	Ⓜ
Ⓝ	Ⓝ	Ⓝ	Ⓝ	Ⓝ	Ⓝ
Ⓞ	Ⓞ	Ⓞ	Ⓞ	Ⓞ	Ⓞ
Ⓟ	Ⓟ	Ⓟ	Ⓟ	Ⓟ	Ⓟ
Ⓠ	Ⓠ	Ⓠ	Ⓠ	Ⓠ	Ⓠ
Ⓡ	Ⓡ	Ⓡ	Ⓡ	Ⓡ	Ⓡ
Ⓢ	Ⓢ	Ⓢ	Ⓢ	Ⓢ	Ⓢ
Ⓣ	Ⓣ	Ⓣ	Ⓣ	Ⓣ	Ⓣ
Ⓤ	Ⓤ	Ⓤ	Ⓤ	Ⓤ	Ⓤ
Ⓥ	Ⓥ	Ⓥ	Ⓥ	Ⓥ	Ⓥ
Ⓦ	Ⓦ	Ⓦ	Ⓦ	Ⓦ	Ⓦ
Ⓧ	Ⓧ	Ⓧ	Ⓧ	Ⓧ	Ⓧ
Ⓨ	Ⓨ	Ⓨ	Ⓨ	Ⓨ	Ⓨ
Ⓩ	Ⓩ	Ⓩ	Ⓩ	Ⓩ	Ⓩ

Section ①   Start with number 1 for each new section.
If a section has fewer questions than answer spaces, leave the extra answer spaces blank.

1. Ⓐ Ⓑ Ⓒ Ⓓ Ⓔ
2. Ⓐ Ⓑ Ⓒ Ⓓ Ⓔ
3. Ⓐ Ⓑ Ⓒ Ⓓ Ⓔ
4. Ⓐ Ⓑ Ⓒ Ⓓ Ⓔ
5. Ⓐ Ⓑ Ⓒ Ⓓ Ⓔ
6. Ⓐ Ⓑ Ⓒ Ⓓ Ⓔ
7. Ⓐ Ⓑ Ⓒ Ⓓ Ⓔ
8. Ⓐ Ⓑ Ⓒ Ⓓ Ⓔ
9. Ⓐ Ⓑ Ⓒ Ⓓ Ⓔ
10. Ⓐ Ⓑ Ⓒ Ⓓ Ⓔ
11. Ⓐ Ⓑ Ⓒ Ⓓ Ⓔ
12. Ⓐ Ⓑ Ⓒ Ⓓ Ⓔ
13. Ⓐ Ⓑ Ⓒ Ⓓ Ⓔ
14. Ⓐ Ⓑ Ⓒ Ⓓ Ⓔ
15. Ⓐ Ⓑ Ⓒ Ⓓ Ⓔ
16. Ⓐ Ⓑ Ⓒ Ⓓ Ⓔ
17. Ⓐ Ⓑ Ⓒ Ⓓ Ⓔ
18. Ⓐ Ⓑ Ⓒ Ⓓ Ⓔ
19. Ⓐ Ⓑ Ⓒ Ⓓ Ⓔ
20. Ⓐ Ⓑ Ⓒ Ⓓ Ⓔ
21. Ⓐ Ⓑ Ⓒ Ⓓ Ⓔ
22. Ⓐ Ⓑ Ⓒ Ⓓ Ⓔ
23. Ⓐ Ⓑ Ⓒ Ⓓ Ⓔ
24. Ⓐ Ⓑ Ⓒ Ⓓ Ⓔ
25. Ⓐ Ⓑ Ⓒ Ⓓ Ⓔ
26. Ⓐ Ⓑ Ⓒ Ⓓ Ⓔ
27. Ⓐ Ⓑ Ⓒ Ⓓ Ⓔ
28. Ⓐ Ⓑ Ⓒ Ⓓ Ⓔ
29. Ⓐ Ⓑ Ⓒ Ⓓ Ⓔ
30. Ⓐ Ⓑ Ⓒ Ⓓ Ⓔ

31. Ⓐ Ⓑ Ⓒ Ⓓ Ⓔ
32. Ⓐ Ⓑ Ⓒ Ⓓ Ⓔ
33. Ⓐ Ⓑ Ⓒ Ⓓ Ⓔ
34. Ⓐ Ⓑ Ⓒ Ⓓ Ⓔ
35. Ⓐ Ⓑ Ⓒ Ⓓ Ⓔ
36. Ⓐ Ⓑ Ⓒ Ⓓ Ⓔ
37. Ⓐ Ⓑ Ⓒ Ⓓ Ⓔ
38. Ⓐ Ⓑ Ⓒ Ⓓ Ⓔ
39. Ⓐ Ⓑ Ⓒ Ⓓ Ⓔ
40. Ⓐ Ⓑ Ⓒ Ⓓ Ⓔ
41. Ⓐ Ⓑ Ⓒ Ⓓ Ⓔ
42. Ⓐ Ⓑ Ⓒ Ⓓ Ⓔ
43. Ⓐ Ⓑ Ⓒ Ⓓ Ⓔ
44. Ⓐ Ⓑ Ⓒ Ⓓ Ⓔ
45. Ⓐ Ⓑ Ⓒ Ⓓ Ⓔ
46. Ⓐ Ⓑ Ⓒ Ⓓ Ⓔ
47. Ⓐ Ⓑ Ⓒ Ⓓ Ⓔ
48. Ⓐ Ⓑ Ⓒ Ⓓ Ⓔ
49. Ⓐ Ⓑ Ⓒ Ⓓ Ⓔ
50. Ⓐ Ⓑ Ⓒ Ⓓ Ⓔ
51. Ⓐ Ⓑ Ⓒ Ⓓ Ⓔ
52. Ⓐ Ⓑ Ⓒ Ⓓ Ⓔ
53. Ⓐ Ⓑ Ⓒ Ⓓ Ⓔ
54. Ⓐ Ⓑ Ⓒ Ⓓ Ⓔ
55. Ⓐ Ⓑ Ⓒ Ⓓ Ⓔ
56. Ⓐ Ⓑ Ⓒ Ⓓ Ⓔ
57. Ⓐ Ⓑ Ⓒ Ⓓ Ⓔ
58. Ⓐ Ⓑ Ⓒ Ⓓ Ⓔ
59. Ⓐ Ⓑ Ⓒ Ⓓ Ⓔ
60. Ⓐ Ⓑ Ⓒ Ⓓ Ⓔ

61. Ⓐ Ⓑ Ⓒ Ⓓ Ⓔ
62. Ⓐ Ⓑ Ⓒ Ⓓ Ⓔ
63. Ⓐ Ⓑ Ⓒ Ⓓ Ⓔ
64. Ⓐ Ⓑ Ⓒ Ⓓ Ⓔ
65. Ⓐ Ⓑ Ⓒ Ⓓ Ⓔ
66. Ⓐ Ⓑ Ⓒ Ⓓ Ⓔ
67. Ⓐ Ⓑ Ⓒ Ⓓ Ⓔ
68. Ⓐ Ⓑ Ⓒ Ⓓ Ⓔ
69. Ⓐ Ⓑ Ⓒ Ⓓ Ⓔ
70. Ⓐ Ⓑ Ⓒ Ⓓ Ⓔ
71. Ⓐ Ⓑ Ⓒ Ⓓ Ⓔ
72. Ⓐ Ⓑ Ⓒ Ⓓ Ⓔ
73. Ⓐ Ⓑ Ⓒ Ⓓ Ⓔ
74. Ⓐ Ⓑ Ⓒ Ⓓ Ⓔ
75. Ⓐ Ⓑ Ⓒ Ⓓ Ⓔ
76. Ⓐ Ⓑ Ⓒ Ⓓ Ⓔ
77. Ⓐ Ⓑ Ⓒ Ⓓ Ⓔ
78. Ⓐ Ⓑ Ⓒ Ⓓ Ⓔ
79. Ⓐ Ⓑ Ⓒ Ⓓ Ⓔ
80. Ⓐ Ⓑ Ⓒ Ⓓ Ⓔ
81. Ⓐ Ⓑ Ⓒ Ⓓ Ⓔ
82. Ⓐ Ⓑ Ⓒ Ⓓ Ⓔ
83. Ⓐ Ⓑ Ⓒ Ⓓ Ⓔ
84. Ⓐ Ⓑ Ⓒ Ⓓ Ⓔ
85. Ⓐ Ⓑ Ⓒ Ⓓ Ⓔ
86. Ⓐ Ⓑ Ⓒ Ⓓ Ⓔ
87. Ⓐ Ⓑ Ⓒ Ⓓ Ⓔ
88. Ⓐ Ⓑ Ⓒ Ⓓ Ⓔ
89. Ⓐ Ⓑ Ⓒ Ⓓ Ⓔ
90. Ⓐ Ⓑ Ⓒ Ⓓ Ⓔ

91. Ⓐ Ⓑ Ⓒ Ⓓ Ⓔ
92. Ⓐ Ⓑ Ⓒ Ⓓ Ⓔ
93. Ⓐ Ⓑ Ⓒ Ⓓ Ⓔ
94. Ⓐ Ⓑ Ⓒ Ⓓ Ⓔ
95. Ⓐ Ⓑ Ⓒ Ⓓ Ⓔ
96. Ⓐ Ⓑ Ⓒ Ⓓ Ⓔ
97. Ⓐ Ⓑ Ⓒ Ⓓ Ⓔ
98. Ⓐ Ⓑ Ⓒ Ⓓ Ⓔ
99. Ⓐ Ⓑ Ⓒ Ⓓ Ⓔ
100. Ⓐ Ⓑ Ⓒ Ⓓ Ⓔ
101. Ⓐ Ⓑ Ⓒ Ⓓ Ⓔ
102. Ⓐ Ⓑ Ⓒ Ⓓ Ⓔ
103. Ⓐ Ⓑ Ⓒ Ⓓ Ⓔ
104. Ⓐ Ⓑ Ⓒ Ⓓ Ⓔ
105. Ⓐ Ⓑ Ⓒ Ⓓ Ⓔ
106. Ⓐ Ⓑ Ⓒ Ⓓ Ⓔ
107. Ⓐ Ⓑ Ⓒ Ⓓ Ⓔ
108. Ⓐ Ⓑ Ⓒ Ⓓ Ⓔ
109. Ⓐ Ⓑ Ⓒ Ⓓ Ⓔ
110. Ⓐ Ⓑ Ⓒ Ⓓ Ⓔ
111. Ⓐ Ⓑ Ⓒ Ⓓ Ⓔ
112. Ⓐ Ⓑ Ⓒ Ⓓ Ⓔ
113. Ⓐ Ⓑ Ⓒ Ⓓ Ⓔ
114. Ⓐ Ⓑ Ⓒ Ⓓ Ⓔ
115. Ⓐ Ⓑ Ⓒ Ⓓ Ⓔ
116. Ⓐ Ⓑ Ⓒ Ⓓ Ⓔ
117. Ⓐ Ⓑ Ⓒ Ⓓ Ⓔ
118. Ⓐ Ⓑ Ⓒ Ⓓ Ⓔ
119. Ⓐ Ⓑ Ⓒ Ⓓ Ⓔ
120. Ⓐ Ⓑ Ⓒ Ⓓ Ⓔ

NOTES

NOTES

NOTES

NOTES

AP Exams

Cracking the AP — Biology,
2004–2005 Edition
0-375-76393-7 • $18.00

Cracking the AP — Calculus AB & BC,
2004–2005 Edition
0-375-76381-3 • $19.00

Cracking the AP — Chemistry,
2004–2005 Edition
0-375-76382-1• $18.00

Cracking the AP — Computer Science AB & BC, 2004-2005 Edition
0-375-76383-X • $19.00

Cracking the AP — Economics (Macro & Micro), 2004-2005 Edition
0-375-76384-8 • $18.00

Cracking the AP — English Literature,
2004–2005 Edition
0-375-76385-6 • $18.00

Cracking the AP — European History,
2004–2005 Edition
0-375-76386-4 • $18.00

Cracking the AP — Physics B & C,
2004–2005 Edition
0-375-76387-2 • $19.00

Cracking the AP — Psychology,
2004–2005 Edition
0-375-76388-0 • $18.00

Cracking the AP — Spanish,
2004–2005 Edition
0-375-76389-9 • $18.00

Cracking the AP — Statistics,
2004–2005 Edition
0-375-76390-2 • $19.00

Cracking the AP — U.S. Government and Politics, 2004–2005 Edition
0-375-76391-0 • $18.00

Cracking the AP — U.S. History,
2004–2005 Edition
0-375-76392-9 • $18.00

Cracking the AP — World History,
2004–2005 Edition
0-375-76380-5 • $18.00

SAT II Exams

Cracking the SAT II: Biology,
2003-2004 Edition
0-375-76294-9 • $18.00

Cracking the SAT II: Chemistry,
2003-2004 Edition
0-375-76296-5 • $17.00

Cracking the SAT II: French,
2003-2004 Edition
0-375-76295-7 • $17.00

Cracking the SAT II: Writing & Literature,
2003-2004 Edition
0-375-76301-5 • $17.00

Cracking the SAT II: Math,
2003-2004 Edition
0-375-76298-1 • $18.00

Cracking the SAT II: Physics,
2003-2004 Edition
0-375-76299-X • $18.00

Cracking the SAT II: Spanish,
2003-2004 Edition
0-375-76300-7 • $17.00

Cracking the SAT II: U.S. & World History,
2003-2004 Edition
0-375-76297-3 • $18.00

Available at Bookstores Everywhere
PrincetonReview.com